Ian Ridley is a football columnist with the *Observer* and a scriptwriter. He has written five books, including *Season in the Cold: A Journey Through English Football*, *Cantona: The Red and the Black* and *Addicted*, the bestselling autobiography of Tony Adams, which was shortlisted for the 1999 William Hill Sports Book of the Year Award.

tales from the boot camps

steve claridge

with Ian Ridley

ORION

An Orion paperback
First published in Great Britain by Victor Gollancz in 1997
This paperback edition published in 2000 by Indigo,
an imprint of Orion Books Ltd,
Orion House, 5 Upper St Martin's Lane, London WC2H 9EA

Third impression 2002

A CIP catalogue record for this book is available from the
British Library.

0 575 40268 7

Printed and bound in Great Britain by
The Guernsey Press Co. Ltd, Guernsey, C.I.

Contents

Introduction

We didn't know who this young lad playing up front was but we liked what we saw. He held the ball up neatly with as cushioned a first touch as you were likely to get at this level of the game. He turned cleverly and took on defenders bravely. Above all, he had a heart like a lion, the arms and legs pumping and cheeks puffing as he gamely ran for every lost cause. The socks rolled down to the ankles were a symbolic trademark of the all-action, value-for-money style. Football fans immediately take to such players and forgive them much.

It was Saturday October 12, 1985, and a group of seven or eight of us anoraked nutters from the London supporters branch of Weymouth Football Club, my home-town team, had come to Wycombe Wanderers' sloping old town-centre ground at Loakes Park for this match in the Alliance Premier League – now the Nationwide Football Conference – unaware of the coltish figure gambolling enthusiastically in the cause of the Terras; so nicknamed from their colours of terra cotta and blue. The opportunities to purchase the *Dorset Evening Echo* in the environs of the capital were, after all, distinctly limited and we were a few days behind on club news.

A few terrace enquiries to our acquaintances in the travelling band of supporters still resident in our old home town elicited a sketchy biography. Aged 19, name of Steve Claridge. A striker on loan from Bournemouth. Had been an apprentice at Portsmouth, where he was born, before brief spells at Basingstoke and Fareham Town of the Southern League. Weymouth were seemingly flush with money at the time and about to move to a new stadium on the edge of town, having sold for £3 million (to a supermarket chain, of course) their, to me, memory-filled but deteriorating old Recreation Ground, hemmed in by the harbour, gasworks, disused railway line and Department of Health and Social Security offices. As one of the top non-league sides in the country then, having finished runners-up to Altrincham in the Alliance's inaugural season, they were pushing for the Football League now that automatic promotion had finally been established, and were willing to pay hard-up Bournemouth £10,000 for Claridge. The striker was apparently dubious, however, about dropping out of the full-time professional game to become a semi-pro.

Elated after a 3–0 victory, some achievement at Wycombe in those days and a result we Weymouth exiles pine for in these depressing Southern League times, we adjourned to the bar. This was as much a pleasure of non-league football as the day itself, which was generally free of the rigours involved in supporting bigger clubs. There was no need to get there three hours before kick-off to pay £5 for a parking space; rather an hour early, parking in a safe, neighbouring street, then through a queueless turnstile for a pre-match drink in the social club. And there was no claustrophobic, seated discomfort, but instead ease of movement away from the noisy, abusive and ill-informed nuisance, and unhindered access to the tea bar at half-time. Even if the standard of football and the feeling of isolated inferiority for the non-league supporter did grate a little, the match was always fiercely fought, the heart full of optimism that better days were ahead.

In the bar, post-match, you got to chat to the players, too. Try that at Old Trafford: 'Tell me, Eric, what was going through your mind that night at Selhurst Park? I have this theory . . .' They probably hated it, having to stand there and endure the ignorance and humour the punters, but Claridge didn't that day. His permanently grinning face seemed to be enjoying all the compliments flying his way as he listened closely without interrupting. What a nice, simple, unaffected lad, you couldn't fail to conclude.

Come and sign for us, I instructed him. You'll be better off playing in front of 1000 people at this level rather than a couple of hundred for Bournemouth reserves. More important from your own point of view, you'll score 20 goals a season and get noticed in the game a lot more readily. Hadn't Graham Roberts and Andy Townsend gone on from Weymouth to Tottenham and Southampton respectively?

He seemed to take it in. Anyway, within a month he had signed for Weymouth to begin two and half years that yielded 57 goals from 149 appearances and became a part of probably the best team they had had in their history. Soon there would be Peter Guthrie, later to join Spurs for £100,000, in goal; Tommy Jones, who moved on to Aberdeen, Swindon and Reading, in midfield; up front, flanking Claridge, the skilful Tony Agana, who went on to Watford, Sheffield United and Notts County. Another contemporary was the centre-back

Shaun Teale, who moved on to Bournemouth then Aston Villa.

In Claridge's third and what proved to be final season Weymouth should have won the Conference. Now installed in the new concrete and characterless Wessex Stadium on an industrial estate near the Radipole Lake bird sanctuary, they won their first five games, including a 3–0 demolition of relegated Lincoln City in front of 3600, and come November they still led the league. Then a spring burst from underground in the middle of the field making the pitch unplayable throughout the winter months. Home games had to be taken to a variety of grounds in the South-West and the momentum built up was lost. They finished well adrift, Lincoln returning to the Football League, and the chance was gone on the winds around the Wessex. There was a darkness on the edge of town from which the club, a decade later, has yet to recover.

Steve knew too that the moment had passed. Now 21, he felt he had to move on if he was going to make headway in his career. The way he was allowed to leave said much about how Weymouth was being run at the time and how the club was slipping. Due to an oversight, he never received a letter retaining his services and, out of contract, was permitted a free transfer. After considering offers from Barnet and Wycombe Wanderers, he could not resist the approach of Steve Coppell on behalf of Crystal Palace, then a Second Division club, that summer of 1988. His desire to be a league footballer seemed about to be fulfilled.

I too drifted away from Weymouth, my work having changed at my then employers the *Guardian* from a production journalist to a football reporter, which demanded duty and attention elsewhere on the Saturday afternoons that had previously been days off. I always kept half an eye on the Claridge career, though, as you do with a soft-spot favourite player who has meant something to you in some way.

Crystal Palace was not for him; Mark Bright and Ian Wright barred the way and he lasted only a few months – beset by fines for frequent late arrivals at training and reserve team matches that anyone who has negotiated the south London streets around Selhurst Park will sympathize with – before fetching up at then Third Division Aldershot, who paid £14,000 for him. After eighteen months slumming it at an ailing, paint-peeling club going out of business, he was sold to Cambridge United for a desperately needed £75,000 that kept the wolf from Aldershot's door for a while longer. Superficially, all seemed well there as John Beck's on-a-roll team threatened the top flight of the game with their ugly and brash brand of long-ball football, though, unhappy, Steve was in and out of the team. In all, he spent four years at the Abbey Stadium, interrupted only by a few months at Luton, more miserable even than his first spell at Cambridge as he failed to settle and played, according to him, the worst football of his career.

Then, in January of 1994, the ebullient Barry Fry bought him for Birmingham City for £350,000 to signal what was an exciting, never-a-dull-moment spell at the hub of the publicity-hungry, soap-opera regime of David Sullivan and Karren Brady, soft-porn publisher and his right-hand woman. Brum went down to the new Second Division in his first season but bounced back into the First the next with Steve becoming the first City player in the two decades since Trevor Francis to score 20 league goals in a season.

It was during this period that I encountered him again, after covering an FA Cup tie against Liverpool at St Andrews, from which Birmingham emerged with a creditable 0–0 draw. We nodded recognition when he came up to the press box to do a radio interview, then chatted for a few minutes about the good and bad old days at Weymouth, whose current plight saddened him.

A year later, in the winter of 1996, Birmingham reached a Coca-Cola Cup semi-final against Leeds United and I recalled Claridge's undulating – 'much-travelled' being the standard description – career when I was seeking a subject for an article as preview to the game for the *Independent on Sunday*, whose football correspondent I had become. I rang Barry Fry, always one of the football writer's standard contacts, who gave me Steve's phone number in Luton, where he had bought a house during his brief career with the town's team. It was a start of three days trying to nail down a jelly, a period that said much about his hither-and-thither, at times chaotic, way of life, but eventually we coincided. It became one of the best interviews a footballer had ever given me and one that spoke volumes about the life of a professional in the modern era, a time which has seen the game explode with money, intrigue and interest at the top level but remain villagey in its everybody-knows-everybody-else's-business elsewhere.

It was midday on the Wednesday before publication when, having been trying the number every half hour for two days, I finally got through. ''Ullo?' enquired a voice that knew not what time of day it was and whose bleary eyes and dishevelled, cropped hair you could picture without the need for any new-fangled videophone. He was sorry, he said, but he had been staying with his parents in Portsmouth, between where, Birmingham and Luton, he commuted. It sounded like a central England triangle into which he frequently disappeared. Meet me in the gym at Birmingham University tomorrow at two o'clock, was the best sense I could make of him.

I duly turned up to watch training, snow having driven City indoors to the university. 'We'll do it at the end,' he said even before I could finish my 'Hello, Steve. I was just wondering . . .' introduction. At the end, he was in a rush to get away. 'Bit of business to sort out with Karren down at the club,' he offered by way of explanation, scribbling down an almost illegible address on a scrap of paper. 'Come to my house in Luton at eight tonight.' By now I was beginning to get a feel for his character. 'You will definitely be there, won't you?' I asked pleadingly.

He was and the mad motorway dashing proved worth it. I knew I was on promising ground when I began with a bit of idle chatter about the training that afternoon and mentioned that Birmingham was the largest campus university in Britain with some 14,000 people. 'Only a few more than we've got, then,' he replied, talking of his club, who under Barry Fry's manic transfer dealing policy that season would break the English league record for the number of players used: 46. The previous record was held by, um, Barry Fry and Birmingham, with 43 the previous season. In all, more than 100 players either came or went under 'Baz' – 63 of them incomers – though Claridge mostly remained one of the favoured, regularly chosen few.

With complete candour and charming wit, Claridge trawled through his amazing life and career, beginning with his adoption as a baby of six weeks, the heart defect discovered when he was 12 and now controlled by medication, his jobs as a gardener (sacked three times) before making it to some of football's twilight zones and boot camps, forever on the fringe of the big time. 'People have said to me: "Don't you regret not doing any better?"' he told me. 'But I have been in some unbelievable scrapes, met some great characters, played at some mad clubs and I wouldn't swop any of it. It's just been a barrel of laughs right from the off.'

There was Aldershot and training on dog-fouled public parks, where the central midfield player Giorgio Mazzon had a disabled sticker on his car and Claridge sold fruit and vegetables from the back of his own car to the other players to supplement his income. At Cambridge United there was the hard-line regime and the rigorous routines of John Beck with his ninety-minute one-touch training sessions, punishing fitness programmes and bizarre philosophies. And Birmingham, where pre-season training on Dunstable Downs resembled a scene out of the film *Zulu* ('There were that many of us coming over the hills') and he played with 23 different striking partners in his two-year spell for a club he loved for its passionate, working-class supporters, with whom he had much empathy. What summed up his often haphazard, unplanned life, he said (proposing an interlude in the interview at this point while I took him for a fish-and-chip supper), was the fact that he had only ever bought two houses, one in Wimbledon when he joined Crystal Palace and the other in Luton. The first he never lived in, the second was while at a club with whom he spent only three months. 'He was a difficult player to coach,' his manager there, David Pleat, told me. 'So unpredictable.'

It all combined to form not only a portrait of one of the game's most colourful characters but also a fascinating, unseen picture of life in the lower divisions, one to which my 1200 words for my newspaper could barely do justice. There had been, after all, spells with eleven different clubs, and in fourteen seasons as a professional he had only ever five times finished a season at the same club with which he began it. One of his previous managers, Harry Redknapp at Bournemouth, now of West Ham, had once said that you could write a book about Steve, familiar as Harry was with all the tales surrounding him – and indeed party to some himself. The thought occurred that Steve deserved one.

When I put the idea to him he readily agreed, wanting to give people an insight into the life and times of a pro beyond the back-page headlines and behind the slick Sky television images that attend the multi-million-pound Premiership. 'There may not be much money in it,' I warned him. 'That's all right. I'm not Bryan Robson, am I?' he replied. Not even the best-paid, or most fêted or famous, have tales like his, I believe, however. Neither can many tell them with the affectionate but challenging frankness Claridge ventured. Some of this must be far-fetched, I thought, at points during our conversations. I decided to speak to a collection of former colleagues and managers. They confirmed it all, even adding anecdotes of their own.

He was determined to be rigorously honest about himself as well as others, and volunteered readily what football's grapevine had known for some time but was barely ready or able to print: that a serious gambling addiction had coursed its draining and depressing route through his career. It had, he told me, cost him at least £300,000 in the past ten years and had got him into some serious and miserable situations. He once considered robbing a supermarket, so desperate was he. 'I've earned a fortune in my career and sometimes I haven't had a pot to piss in,' he said as we talked at his home, where the kitchen had bare boards, a patio was half-built and the living room was furnished sparingly as he divided his time between this footballing base and the warmer family farm at Titchfield Common, near Portsmouth, where his wife Mandy, whom he married in June of 1995, also lives and indulges his career patiently.

Shortly after the unsuccessful Coca-Cola Cup semi-final against Leeds, in which he was still cheered by Birmingham fans despite missing a penalty, an acrimonious transfer around his thirtieth birthday took him to Leicester City for £1.2 million. With the gambling now under control – though he admits that, as with any addictive illness, he may only be in remission and may need help finally to overcome – and following a thyroid problem that debilitated him for nearly three months, he played a crucial part in Leicester's late run to the First Division play-offs.

Then came the crowning moment in front of a full-to-capacity, sun-drenched Wembley Stadium. Barely a minute of extra time remained against Crystal Palace, who had let him go almost eight years previously, with the score locked at 1–1 when the ball sat up invitingly for him on the edge of the Palace penalty area. 'It was winking at me,' he told me later. In an instant he had despatched it past Nigel Martyn for the goal that took Leicester up. Joyously, all exhaustion forgotten, he set off on the celebratory run of his life as the blue-and-white half of the faded but still grand old stadium erupted. The venue of legends had a new one; well, at least Leicester did. It meant millions of pounds to the club and a chance to dream anew for their fans.

When he had calmed down – about a month later – Steve told me: 'Just think. That goal means we are going to play at Old Trafford against Manchester United while Palace will be going to Grimsby.' It duly happened. The fixture computer arranged for the two clubs to take their widely disparate journeys on the same day in the November of the next season. (Poor Grimsby, incidentally; they seem to have replaced Oldham as the epitome of the unglamorous. So disparagingly referred to, so often, are they that their name seems unofficially to have been changed to Thelikesofgrimsby.)

After all this time, Steve's fifteenth season as a professional was going to see him play in the Premiership, a fitting reward for a worthy, persistent career, one that has improved with maturity. He was to go on and, astonishingly, score the winning goal in the Coca-Cola Cup Final against Middlesbrough. 'The way it's gone I'll sign for Manchester

United on my fortieth birthday,' he said. Perhaps by then his gambling will be long gone and he will have banked rather than squandered the financial rewards that came his way so late in his career. He deserved his move to Wolverhampton Wanderers before going home to Portsmouth.

All the while, Steve has been seen as just a journeyman player, one who could do a good job for a lower division team. Seen, too, as football's Just William with his collection of old muddied boots that he insists should not be cleaned (make that football's Imelda Marcos), socks at half-mast that have most referees telling him to pull them up or face a booking, and a spiky-haired, scruffy appearance allied to a mischievous, winning smile. Danny Baker's Radio 5 programme once noted that he resembled the American comedienne, Joan Rivers.

He has always been realistic about his abilities but slightly resents the image, believing it overlooks the fact that he has a touch of quality too. But for the early immaturity, notorious timekeeping and penchant for punting – drawbacks, granted – he might have made the top sooner, though things in life often happen when they are meant to, when people are properly equipped to handle them. Perhaps Steve would not have been ready before.

'I did come close to signing him a couple of years ago,' one Premiership manager told me, 'but he's mad, isn't he?' Yes and no. Unconventional and erratic maybe, but an honest, skilful player who is the lifeblood of English professional football; a salt of the earth who deserved his day in the sun. Steve Claridge's story, as unusual as it may be, is an example of why the domestic game is such a rich, deep and diverse competition, one unmatched anywhere in the world. Few anywhere will match the way he tells it, either.

1 A Gambling Man

There was once a football manager who asked a player straight up when they were discussing a possible transfer what his problem was: women, drinking or gambling. 'You may as well tell us, son, because we'll find out soon enough,' he said. These days, with so much time on their hands and money in their pockets, most players are supposed to have at least one expensive, high-profile vice. That has not been my experience, generally, and I can only hold my hands up to my own problem. Most of the managers who have signed me – twelve of them now – have known what my passion was, or became aware of it pretty quickly.

With me, it has always been gambling, mainly on horses, and I know the word has spread about me on the game's bush telegraph. Some have seen me as a bit of a joker, as a character, the gambling a big part of me. It has probably put some managers off signing me, especially at the highest level, although I believe sincerely that I have always given every ounce of me on the field. There, I have always felt in control of myself and my life, doing what I do best. I heard Paul Gascoigne saying once that the pitch was the only place he truly felt free and, although I have never had the same glare of publicity as him, I know exactly what he meant.

Off the field it has been a different matter. The gambling has often controlled me, bringing a misery far beyond the £300,000 or so I reckon I have 'done' during my career. In the past few years, it became such a problem, going up from £10,000 a year to £50,000 as I earned more, that I knew eventually I would have to do something about it. Admitting to it is the first step. I think my friends and family will be staggered by the amounts involved. They have known little beyond the fact that I have always liked a bet. When you are ashamed of something, and I know I haven't done the right thing, you try to keep it hidden. I think it is fair to say that I have been gripped by a gambling addiction.

It started off pretty small-time, as most addictions do. When I was 10 or 11, my dad would let me join in the sociable, small-stakes card schools he used to host on Friday nights on his green baize table for a neighbour, Jim Shanley, and three or four policemen of his acquaintance. It was a bit like the Keystone Kops, really. Sometimes they would get a call-out but would only go if it was an emergency. Otherwise they would say they were tied up. It was all pretty friendly with my dad, who has never been a big gambler, letting me use the small amounts he paid me for helping in his market gardening business. Nobody ever won or lost very much, with perhaps a big 'pot' being £30, but I suppose it gave me a taste.

My first real bet, with neighbour Jim placing it for me, was at the age of 11: £1 on Virginia Wade to win Wimbledon in 1977. She duly romped home at 16–1, with me having skived off school to watch it on television. I also recall winning quite big on Ben Nevis in the 1980 Grand National at 40–1. The Derby was an annual occasion when I used to stay home from school, pretending I had a sore throat. 'Got the Derby throat again, have we?' my mum would ask. With Jim, I also used to do a regular Saturday bet of 36 doubles, trebles and accumulators, my share being £18. It doesn't seem a lot of money now and even then it didn't seem much to me, just a bit of fun and no more, but I suppose to others it would, especially with it being a 14-year-old involved. My dad knew, but he didn't seem to mind. He thought it was just a youthful diversion, I suppose.

During my teenage years I was betting only what I earned off my dad or from selling strawberries by the M27 on days when I bunked off school and tagged along with traders I knew. I suppose I was losing £20 to £30 a week but I was doing nobody any harm – not even myself, I thought. I didn't really get that much of a kick out of it then, and I had no responsibilities.

Neither was it a problem during the apprenticeship I began with Portsmouth at the age of 16. We were far too busy all day and I was scratching a living on £25 a week, having my bed and board provided at home, so the habit could hardly get out of control. I did venture into the bookie's now and then on the days when I should have been on a day-release course at High-bury College. Then came the day when Alan Ball told me I was not going to make it at Portsmouth, fifteen months into something I had so desperately wanted. The club, naturally enough, had taken a dim view of my non-attendance at college, and coupled with timekeeping that sometimes saw me late for training, they were going to release me. When Bally, then the youth team coach, had me in his office to tell me, I was upset. But it was a strange feeling, because at that age of 17 I didn't really understand the consequences. I thought to myself, Well, if I don't make it here, I'll make it somewhere else. It was the same with the money, I always thought I had earning potential and would have enough for my needs, which included betting.

Alan Ball handed me a cheque for £500 as six months' pay-off, and it should have seemed like a huge sum I suppose, but that wasn't how I saw it. It wasn't money to cushion the blow or tide me over until I found a new club or something else to do with my life, as most sensible people might decide; it was just there to be spent. A cheque was no good to me so I just got the club to cash it and I went immediately back to Titchfield to a bookmaker's there. Within three or four hours it had gone on several horses whose names I have long since forgotten. It was my way of dealing with no longer being a professional footballer. I needed some relief and consolation. As I walked out of the shop I can remember feeling a little bit gutted at losing the money. Not gutted enough, obviously.

It's a strange thing, but you lose your grasp on the value of money when you are gambling. If I had spent £500 on clothes, which would have seemed a real extravagance, I might have been concerned. But when you go into a bookmaker's, it feels like being a character in *The Lion, the Witch and the Wardrobe*: you're going through the wardrobe into another world. It's just like a fantasy. Time doesn't matter. You can be in there four hours but it feels like four minutes. There always seems to be something happening, with races going off every few minutes on, nowadays, a whole bank of television screens. If you are a sensible punter, you might pick one horse a day, or back nothing for a couple of weeks, and you can do fairly well. The mugs are those who stay there all day. That's the way it gets you. I became a mug.

Not immediately, though. After I left Portsmouth, when I broke my ankle as an 18-year-old playing for the Hampshire county youth team, I was on the dole for nine months and struggling financially. Only when I got back into the game with Bournemouth, then Weymouth, did I become a regular punter again. I remember at Weymouth getting a tip for a horse called Aquilifer several months before the Cheltenham meeting. One friendly bookmaker – yes, they do exist, though only when you are offering them money – let me have 8–1 ante-post and pay for the bet in instalments. I decided I wanted £1200 on it, and it took me eight weeks to pay it off. I won more than £9000. It's funny how you remember the names of the horses that win for you but only very rarely those many losers.

I do believe that my gambling never affected my football – the way I played that is – but I have to be honest and admit that it has come close to infringing on it on the odd occasion. At Weymouth I once turned up for a game at the appointed time an hour before kick-off but nipped out to get a bet on at a bookie's near their old Recreation Ground (the new one in the town's suburbs later presented a problem). Other players also wanted me to lay their bets, with me becoming a sort of unofficial runner, as was also to happen at subsequent clubs, and I made it back only five or ten minutes before kick-off, having stayed on to watch the race. Everyone knew where I had been. My reputation was developing. I earned a roasting from the manager Brian Godfrey as I changed hurriedly, but I still went out and performed up to standard, I believe.

After Weymouth, in my brief spell at Crystal Palace, I never really had any time for gambling. I was preoccupied with trying to get a league career off the ground so the bookie's took a back seat. Also, I seemed to be spending the whole day in the car travelling up from Portsmouth; that and losing most of my wages in fines for turning up late for training and reserve games. I didn't have a great deal to spend. Most of my pay was taken out at source.

The problem really started to kick in at Aldershot. I was earning £300 a week and getting paid monthly. Or was supposed to be. Sometimes, when the club was going through a rough patch, we didn't get paid. When I did get paid, it would usually disappear in two weeks, but that was good for me, with all the craziness that would come later. The rest of the time I was living on the money I got from the fruit and veg stall I set up after training from the back of my car, selling produce to the other players. As soon as that was done, I would get straight down the bookies'. There were three or four right next to the ground around the High Street and we never trained beyond 12.30 p.m., so I could generally be there when racing started.

I can remember once racing to get a bet on and in my rush having an accident in my car, one of at least eight I have had (though not all of them have been gambling-induced). It was my fault, pure and simple. I drove into a lorry, and I wasn't going to argue with the lorry driver, who was a burly bloke. At first I was angry and impatient as I realized the race time was approaching and I wasn't going to get the £1500 on the horse. Then later when I heard it had lost, I was grateful. The bill for the damage to his lorry and my car came to £800, so I figured that crash saved me £700. The driver couldn't believe I was so willing to pay and seemed a bit baffled by the smile on my face when I took the money round to his house a few nights later.

I also got to know professional gamblers barred from betting at racetracks by the on-course bookies, as you do when you become familiar with the twilight world of bookmakers' shops. One asked me to run an errand for him, saying he would make it worth my while. As an unknown face – which is not difficult when you play for Aldershot – I went up to Windsor races one night just to put a bet on one race. I was in and out in ten minutes. The horse won and I left there with £13,000 in cash. My share as courier was a welcome and easy £1300.

It was after my move to Cambridge United, when my wages increased substantially, that things got really bad. The football wasn't going well. I was stuck in a contract for two and a half years with a manager, John Beck, who I despised and who didn't like me. He wasn't going to let me go and it became a personal thing. The squad was also thin but no matter what, I wasn't going to get in the side regularly. He was very stubborn. I turned to regular, expensive punting as a release from the gloom of it all. In a betting shop I felt I could escape from the pressures of the football and there was no one there to bawl me out.

At first Cambridge put me in a hotel, then digs, but after a while the club decided they had paid for me for long enough and told me to find a place of my own. The problem was that I was gambling all my money and didn't have enough money to find anywhere to live. I didn't want to waste it on accommodation and the non-essentials. I even slept in the Cortina I then had on loan from Dad for a couple of nights. Now I was doing all my money a month in advance. I had a mortgage on a place I had bought in Wimbledon when I was at Crystal Palace and that was being deducted at source, along with some bills, because I would just let arrears pile up if I had been given the money. One month I remember having £3.19 left after deductions. I did try to change some cheques with the club but mostly they bounced.

One of the directors had a flat that he let to one of the players, who had to get out when the director decided to sell the place. The player concerned and I had a spare key made and I moved in as a squatter. They knew someone was in there but nobody cottoned on it was me. It went on for about two months until one day I was sleeping on a mattress and heard the front door go. It was an estate agent showing someone around the place. I had to jump up, hurriedly put on my trousers, stash the mattress and climb out through the window carrying a sleeping bag.

The gambling had a grip on me. Because Beck was often having me working out after the others as punishment for not giving in to his methods, I was going to training with my money in my tracksuit trousers, in a zipped-up pocket, so that I could get straight off to the bookie's. Then it was hotfoot to get the money on, not bothering to shower or change until I got home at about six o'clock.

Once, I don't know how – probably rare winnings from the previous day – I had managed to get together £1000 and put it all on a horse called Presidanimich at even money. I was so sure about it. And it was only a four-horse race. Naturally, it didn't win, though I remember this loser's name because of what happened three days later; it romped home at 4–1, though it may have been finishing the previous race. Needless to say I didn't have anything on it. Just another hard-luck story.

Another time, I was really flush, probably after getting a chunk of a new season's signing-on fee, and I did the whole £4500 in one afternoon. On the way back to wherever I was living at the time, I ran out of petrol. It had never occurred to me all day just to save a tenner for petrol. If I had cash I was going to spend it on gambling and no other consideration was going to get in the way. I was just caught up in it, couldn't control it. I still didn't really realize what was happening to me. It all seemed a natural way of life, just a bit of fun, though in hindsight fun was the last thing it was.

I actually ended up running a book, working behind the counter for a bookmaker who became a friend, though these days, with hindsight gained from bitter experience, I think the two things do not go together. I used to use a little independent in a village near Cambridge. After I moved out of the squat, the owner of this bookmaker's took pity on me – I suppose I had been good business for him – and let me sleep on his sofa. Later he moved into a bigger house and I had my own room. I rarely had a bet with him after that because it would have made life difficult for both of us. I suppose I thought that working behind the counter taking money rather than being the one handing it over might curb my habit but often I would go out of the shop and bet somewhere else. He still ended up owing me about £13,000 but there was no way he could pay. Both of us were broke. I suppose he was too soft on people. He let too many people bet on credit and kept extending it but they could never pay him back. In the end, we fell out, as people always do over money.

Some nights I used to stay in his shop on my own till ten, just listening to the commentaries on the night racing, having put a bet on somewhere else during the day. I was in the environment all the time. I suppose I used to get kicks out of watching it all. Whether he could let me have any money he owed me for a bet depended on how the punters did, so I would want them to lose all afternoon. If one or two big players came in, I used to enjoy seeing them lose.

I was really down around this time, one particular incident summing up my state of mind. I remember going to a supermarket to do a bit of shopping just before they closed one night. At the check-out a young lad had left the till open. I looked at all the money in there. All of a sudden it went through my mind to push him off the chair and run off with the money. Then sense prevailed, but I realized how desperate I had become. I was basically just living for the gambling. I could get by on my wits and hold creditors off for a month or so with promises, but I always needed money for a bet. I never really had any debts for too long a time, or if I did I would pay them back pretty quickly, but then I never had any money to spare. Gambling money was not spare money.

Most people would have seen my transfer to Luton as welcome, even if the town itself is a bit gloomy, as a chance to get myself straight again. With my money going up from £35,000 a year to £95,000 it should have been. Instead, the more I had, the more I gambled. We went on a pre-season tour to Sweden, the first time I had been abroad with a team. There was a big carnival going on in one town where we were staying and they had a racetrack with a trotting meeting scheduled. I was rubbing my hands with anticipation. I offered to run a book for the lads and when they agreed, I really thought I had a good touch here. There were some good gamblers at Luton at that time, out of fifteen lads, probably six good ones.

After we'd had a meal in the grandstand, the racing started. I thought I could not lose. There were about fifteen horses in each race, it was foreign, no one knew anything about them, their form or about trotting. In the first race, Mick Harford picked numbers 1 and 14 in a £10 reverse forecast. Bugger me if they didn't come up. I had to pay Mick about £2500. After that, he was betting £100 a race and would have cleaned me out if he had won. Thankfully he didn't and I got the debt back to about £1500. I thought a week abroad would stop me betting for a while, but I guess I could have found a bet in the Sahara Desert. Camel racing, perhaps.

I did some serious financial damage to myself during the three months I was at Luton. After my accommodation hassles in Cambridge I thought buying a house there would help, but when it came to putting the £10,000 down as a deposit I didn't have it, even though Luton had given me a £35,000 signing-on fee and £10,000 removal expenses. My solicitor had to lend me the money for a few weeks until I could pay it back.

One amazing day during that period I won £16,000. Or rather I didn't. It was just one of those runs when I could do no wrong. Nine out of ten horses came up. The bookie didn't have enough in the till to pay me out though and told me to come back the next day at five o'clock to pick up the winnings. But I got there well before that and started to gamble. At 5 p.m., just as my last horse, called Grey Power I think, was going down, a Securicor man walked through the door with my £16,000. I can still hear the bookie's words to this helmeted figure chained to a metal case: 'It's all right, mate. You can take it back. We won't be needing it any more.'

On other days, I was going to apprentices who earned £35 a week and asking them to go to their cashpoint to get a couple of hundred quid out for me, for which I would then give them a post-dated cheque. I never did have any credit cards or bank cards because I didn't trust myself. I would just have run up bills that I would pay belatedly, thus eating into my future stakes. It was the same with bookmakers' credit and telephone accounts. The one bit of sanity in this whole madness. Anyway, I would never have been able to get a loan because my credit rating was by now so bad.

Cambridge was bad but Luton was worse. At every club I have ever played for, I always felt I earned my money, that I could hold my head up in any town, and that the gambling never intruded on my football. I tried not to have a bet on a Saturday, for example. But Luton remains a big source of regret. I don't know if it was because the gambling was making my life so depressing, but my game was crap as well – even though I still saw it as release – and I knew after four or five games there I had made a mistake.

It was a relief when Cambridge came back for me and Luton agreed to sell because they needed quick money. By now John Beck had been sacked, with Gary Johnson taking over as care-taker, and I had no hesitation in going back. It's strange in football how sometimes you can play badly or create problems and get a move that benefits you. It happens in other jobs, too, I suppose; do it badly and you get moved, sometimes to a better one. It had never been my way, however, and I remain sad at the way I let myself down at Luton. I never want it to happen again.

Actually, I think Gary Johnson wanted me back at the Abbey Stadium because he needed someone to take his bets down to the bookie's, as I had done in my first spell there. This time I kept the house on in Luton and was travelling every day, so the opportunity to gamble was lessened a little, though I was still into it. At the time, I thought I had it under control. In retrospect, I can see this was nothing like the case. I had been doing my money three months in advance, but at least now I wasn't short of cash, becoming the highest-paid player in Cambridge's history, on £110,000 a year. It was just that the bets got bigger as the wages did. I became pretty well known as the team's top punter, and the same reputation followed me to Birmingham, amply demonstrated in one incident before a crucial Third Division promotion match at Brentford. Nearing the ground – I could see Griffin Park as we came along the M4 – racing was on the television in the coach, not that I was paying too much attention, because I was trying to steer clear of the horses on match days, and I was also playing cards with a group of the lads. Then Barry Fry noticed that club owner David Sullivan's horse Circa was running and asked the coach driver to pull off the motorway so he could get some money on. The horse was 20–1 and I thought it had no chance, so I shouted to Baz, with a display of bravado: 'I'll take on that bet.' He wanted £100 each way. Our midfield player Mark Ward also wanted £20 each way, so I took that as well. There was no need to pull off the road, I was now the bookie.

Now as I recall there was also a good horse in the race, which was over seven furlongs, so I was pretty confident. Too confident. I should have taken into account the fact that because the ground was a swamp and racing was only just possible, form can go out of the window. Circa drifted to 25s then 33–1, which only confirmed my confidence rather than had me working out how much I might lose. I felt pleased that I'd taken them on at 20s, considering myself shrewd. When the race started, I was still not taking too much notice as I had a good card hand going. Gradually, though, there was some noise from the front of the coach as the race got under way and I started to take an interest. Circa began to move up the field and, with two furlongs left, was, along with the favourite, the only horse left with a chance. The cheering got louder and louder, with Baz in full-throated 'Go on, my cocker' voice. A furlong out, Circa was on its own and won comfortably. I got off the coach for a big game owing Baz £2500 and Mark Ward £500. Three grand down and I hadn't even had a bet myself. Thank God we won 2–1. At least something lifted my spirits that night.

I had £400 on Birmingham to win promotion at 8–1 that year but the bloke I laid the bet with couldn't afford to pay me, so let me keep on betting with him, in the hope that I would lose it all, as I always had done in the past. Instead, I kept winning and he eventually owed me around £50,000. I had to wait for him to sell some land so I could collect. Doesn't it just always seem to be the way that when you win big, something goes wrong? The only other time I was owed cash was by the bookie in Cambridgeshire, but I suppose I owed him a lot besides money anyway.

The biggest single bet I have ever had was when I was with Birmingham. I went down to Cheltenham for the Gold Cup with my old neighbour Jim Shanley and a couple of other friends and I had £10,000 on Master Oates for the Gold Cup, placing ten different bets at the course of a grand each at odds between 11–4 and 7–2. The horse duly won. That was a great day, but they have been few and far between. Such highlights ignore all the losing days.

And there have been plenty. In the six-year period between 1989 and 1995, taking in Aldershot, Cambridge United (twice), Luton and Birmingham, I had periods of control but quite often I slipped into a seven-day routine of gambling, what with night and Sunday racing. In those days I would usually get up five minutes before I had to leave for training, grab a piece of toast or fruit and have a quick scan of the *Sporting Life*, which I had delivered daily to pick out a few likely nags. Sometimes I would train with £4000 in cash stashed away in my tracksuit trousers, occasionally checking that it was still there when I was running. Fellow players who knew did find it amazing, but I couldn't see what the fuss was about. It was just stake money to me. I never considered what else it might buy.

At the end of training I would get straight into the car in my kit ready for an afternoon's entertainment, if that is the right word. It was more like a necessity, it seemed. If there was no time before my first horse was off, or if one of the other lads fancied a flutter, to be sociable I would find a bookie's in the town near the club I was with. If my first fancy was an hour or two away or I didn't have company, I would head back to Luton. It was always a dash and I would drive home like a maniac, thinking about the rush of that first bet all the time. I have been stopped for speeding several times. I always seemed to have at least ten penalty points on my licence.

There was always a great sense of anticipation. It felt like Christmas every day. There was a strange pleasure in living on the edge, knowing that if you won you could pay the mortgage, the heart fluttering at the thought that if you lost you were in trouble. That was the buzz. There would be no such buzz for a millionaire betting £10. But it wasn't so much the money, although knowing it was money I may not be able to afford somehow heightened the sensation. It was not so much butterflies in my stomach as elephants playing rugby league. The rush then satisfaction as a horse won was the feeling I chased. It was a different rush from the one I have always got playing football. That lasts longer and is more of an excitement at first, followed by a warm glow. Gambling is escapism, a quest for instant gratification. To me football is real life, even if it is escapism for some spectators.

Sometimes I might have done all my money in the first ten

minutes of being in there. I have always been a shit-or-bust gambler rather than one who paced himself. If I fancied a horse, I would often put the lot on. If I lost it, I would have to go out and get some more money by borrowing it, seeing if I had any cash at home or getting a sub from my wages at the club. I never was very good at leaving a bookie's before the last race, like a drinker going home before closing time, I suppose. I was always chasing a winner, trying to get it back. Hoping for that really big hit. But if I got it, I would lose it the next day. I would leave ahead on about two days a week. Which meant five losing days.

My way of gambling began logically but often deteriorated. In the morning, or sometimes once in the shop, I would choose one or two horses that I fancied for the afternoon, usually on form. If they had won over the same distance and ground previously, I would often go for them. I have always had a good memory for horses and, besides, when you are in a bookie's every day, you take notice of what's happening. As the day went on, though, I would often rely on a gut instinct. Then again, I might get desperate and choose things in haste. And repent at leisure.

It was generally the horses that I gambled on, although I would bet on greyhounds now and then. I remember once going with a group of friends to a track on the South Coast having had a tip for a dog. We were told that if it got a clean start from the traps it would win easily, but if it was baulked it lost heart and had little chance. So we hatched a plan. Two of us went to stand by the back straight, the others in front of the grandstand. After the start one group would signal to the other how the dog had started the race. The dog was duly baulked and one of the lads, mingling with the crowd, threw a cuddly toy on to the track. It had the desired effect of sidetracking the dogs, who went for the toy instead of chasing the hare, and disrupting the race, which was then declared void. Thus we got our money back. Naughty but a cash-saver.

If it was a losing day, I would head home at about 6 p.m. on a real downer. It wasn't that I hated myself but I did feel guilty, especially if I phoned my mum who would tell me something like how high her water rates bill was, when I had just done enough money to buy the water company. As for my own bills, I was always able to put people off. Whenever I have been in real trouble, I have always known that I was due a lump sum in a signing-on fee, or such like. Then I would pay it all off, before starting the whole cycle all over again. I preferred to settle things in cash rather than have things going out by standing order or direct debit.

Once home in the evening, my stomach would still be churning and I would just slump in front of the television waiting for it to be ready for food. If I was on a day off, quite often I would not eat but just go to bed. If I had training, I knew I had better put some fuel in so about ten at night I would go out for fish and chips or some other takeaway. It became a miserable existence but I knew there was always racing tomorrow, or a match to look forward to in a day or two. Football has kept me going.

I was insulating myself in the gambling. It became my solace as I escaped the drudgery of life. But although it had once given me some pleasure, by the time I was at Birmingham it was draining me. I knew I had to try and get control. Because I was with a bigger club there was more interest in me and the rumours were flying. People were commiserating with me that I had lost my flat. Newspaper reporters were phoning me up saying they had heard I had lost £50,000 in one day. The stories were being exaggerated – and I have never been bankrupt, as one paper suggested. It was a bit like players going out for a drink and the management hearing that they were legless, but there was a basis of truth, I had to admit. There were not enough hours in the day for me to have done some of the things I was supposed to have done, but twenty-four were enough, I guess.

Early in 1996, as my move to Leicester City was in the wind, I decided that the daily grind had to end. I have come to see that, yes, gambling is an addiction, an illness like alcoholism or drug dependency, but there are differences. You can't drink away £5000 in an afternoon, for example, but you can gamble it. And I am not doing myself as much physical harm, although financially at times it has seemed never ending. That's why I have never had accounts. I know I would still be paying it off in ten years.

It gets depressing when you are earning so much money then having to go to your mum to borrow £100 to get your car taxed but, although the amounts sound staggering, it is not that which is important. Someone who earns £200 a week and gambles £200 a week is going to feel the same things as I did. It is the compulsion, not the money, that needs to be addressed.

In the end it all got to me, this feeling of being tired and listless off the field was too much to continue with. As recovering addicts say, I just got sick and tired of being sick and tired. I think turning 30 helped as well. I realized my career was passing by and that I could have nothing to show for it in a few years' time. Leicester City offered me another chance and Premiership football came my way at last, so I did not want to blow it any more. They were crucial times for me and I got control again.

I still have a bet if I fancy one, I have to admit, but frequently I will go a month without, then have just a small one, like £100 on England to win Euro '96 at 8–1. Another loser, even if

England emerged in credit. I try to be sensible and I honestly don't think I will ever go back to the way I was. I just have one bet and come out. That's enough. It doesn't seem all-important now. I think I have also accepted that I am going to lose overall, that I am doing it for enjoyment rather than trying to make a living from it. I hope I can enjoy it without it taking over my life. It's about striking a happy medium.

I have only rarely had bets on football: Birmingham to win the title that time; Baz picking the right team for a change – that sort of thing. And the time I was hauled before the FA for backing Pompey to win. I would certainly never bet on my own team to lose. I think anyone who has seen me play would understand that immediately. And I have never taken my troubles on to the pitch. That was my respite, my chance to win, no matter the other losing areas of my life.

I didn't have any one day when I thought, That's enough. Just plenty of them. Thankfully I have always had the football and those ninety-minute oases, when I felt powerful, more determined and positive – unlike in a betting shop, when I feel powerless. I guess it must be an addiction because at those times, nothing else in your life matters. Your quality of life goes out of the window. For one or two moments a day, you get oblivion, like a heroin addict, but the rest of the time you are just down. You just build your life around it. You don't realize how it is catching up with you. You get the highs but you go down for longer and longer. At first it's just ten minutes, then twenty or thirty between races, then it's all the time.

There have been several high-profile players whose gambling has been highlighted in recent years: Keith Gillespie, for example, and before that Paul Merson, who is now in recovery. It has not been my experience that it is a widespread problem in the game, though. You get the odd one or two players but it's just like any other group of people in any walk of life. I don't think there are more footballers than dustmen who gamble, for example. It's the same with booze. Like any group of twenty lads who go out together, one or two will have a few too many, whether they are solicitors, road sweepers or footballers. It is just that footballers are more in the public eye than ever. It has always gone on, but now that there's so much attention trained on us, we have to be more responsible.

Neither have I come across drugs very much, though they are supposed to accompany the new-found wealth that footballers have. I have known the odd player smoke marijuana briefly and socially but never anyone who took cocaine. There was a young lad at Aldershot who used to take Ecstasy and deal in it, though he knew better than to offer it to me, but he never amounted to much. You can't when you're throwing up at half-time as he was doing.

I can't understand the need for any stimulation in the form of pills or anything else to play the game. Football itself is enough stimulation for me. A boxer gets rid of his aggression in the ring; you never see an angry boxer out of the ring. The high of playing football is satisfying in itself. Yes, I know that away from the roar of the crowds life can seem mundane and some may seek thrills to replace it, as my gambling has shown, but the game is mood-altering enough for me these days. I have also heard the rumours about certain players in the game, but I don't take them too seriously. The game is a rumour factory, after all. I can only speak about players I have known. I prefer to judge at first hand.

And I can only do something about my own problem anyway. I have lost so much down the years that what's done is done and it is no good worrying about the past. I have accepted that

I am never going to recover my losses, so that helps with trying to control it, in dealing with the present and the future. I realize, too, that it is not something you beat, but it can beat you. I like to think I am honest with myself. I know that if it kicks in again, I will need some help this time from those who have had similar experiences and recovered from them. As for all the other experiences in my footballing career, I am not sure if you ever recover from them . . .

2 Pompey Times

The Nike advert proclaimed that 1966 was a good year for English football; Eric Cantona was born. Another player born that year, in Portsmouth on April 10, may not have gone on to have quite the same impact on the higher levels of the domestic game, but he has certainly made himself a cult figure with each of the clubs he has represented. In his own way, as individual as Eric Daniel Pierre's, Stephen Edward Claridge has been as much of a character, too.

His beginnings were also on the South Coast of his country, and similarly humble. He was adopted at the age of six weeks but has never seen it as anything other than a benefit in his life and speaks warmly of a happy childhood. Anne and Alan Claridge, homely and honest folk, have indeed been remarkably attentive parents to him and his sister Ruth, also adopted through a local Christian charity, and have encouraged and indulged him at every mile of his footballing journey.

The family home was a modest three-up, three-down in the village of Titchfield, near Portsmouth, before Alan acquired a farm nearby for his market gardening business. In this setting overlooking the sea of the Solent, providing a rural respite from the urban clamour of football, Alan went on to build a house next to his own for his son – that was between commuting to Luton to attend to home improvements for his, shall we say, not especially practical child.

If Alan has been the physical provider, Anne is the confidante and carer. The family living room is a shrine to Steve's achievements, his Player of the Year awards from Weymouth, Aldershot and Birmingham City surrounded by snapshots of his career, unopened bottles of man-of-the-match champagne and countless trophies and medals. Pride of place is awarded to that won when Leicester City reached the Premiership via the play-off final against Crystal Palace thanks to 'that goal'.

By his own admission, Steve was no scholar. Sport was redemption for him at school. He excelled locally at cricket, tennis and basketball, but football was his first love. So keen to encourage him were Anne and Alan, who had painted along an inside wall of their garage a touching mural of Stephen Edward being carried by a stork to mark his birth, that they even formed a team to compete in local leagues for him to play in on Sundays.

Undeterred by a heart defect detected at the age of twelve, Steve's ambition through his teenage years, though he followed Liverpool from afar, was to wear the blue and white of Portsmouth. Along with several of English football's town teams, such as Preston and Burnley in the North, Pompey had been one of the game's traditional powers in another era.

Britannia ruled the waves from the port's dockyards and in the time of the great half-back Jimmy Dickinson (764 appearances for the club and 48 for England) the club won the Football League Championship twice, consecutively, in 1949 and 1950. Their record gate of 51,385 for an FA Cup tie against Derby County was established during that period. The club badge of star and crescent with the motto: 'May God be my guide' evoked gentle and prosperous days. Then, in 1961, the maximum wage intervened and Pompey suffered with clubs from centres of similar size, never since able to attract or retain the best players unless, as with Jack Walker and Blackburn Rovers, some wealthy benefactor was nostalgic for his youth. In the early eighties Portsmouth were in the doldrums, having sunk as low as the old Fourth Division.

For a young lad brought up on such stirring history, they remained a big club, however. Though Fratton Park was a terraced, rusting relic of a bygone age when capacity not comfort or safety was the yardstick, to an impressionable kid it was a copious focal point for his dreams. Duly he was spotted by the club kicking about on the local parks and taken on as an associate schoolboy. But then many were. The real achievement was in being offered an apprenticeship. It was the only life Steve wanted, and he enjoyed the camaraderie, but found it hard to adapt to a disciplined way of life. His timekeeping let him down, he did not attend the day-release college course he was supposed to and he did not respond to the part-time youth team coach Ray Hiron. His natural talent, on which he has always relied, was visible, nevertheless.

Trevor Parker, who would go on to be instrumental in Steve's moves later in his career, was a dedicated follower of South Coast football at the time, then manager of Bashley of the Hampshire League before going on to Basingstoke and Poole Town. 'I have been involved with the game for more than thirty years and I don't think there is a player from that area that I haven't known a lot about during that time,' he says.

'I became aware of Steve quite early on. In fact he came to Bashley for a Hampshire Senior Cup tie with a Portsmouth team, and I could see unusual qualities in him. He was ungainly and leggy but with control and touch and he always looked as if he was enjoying himself. I think he would have been sixteen at the time. Even then he was a bit strange, I think. His boots looked three sizes too small, his hair was different to everybody else's and he had this eccentric gait.' (The word 'eccentric' seems to crop up a lot.)

Natural talent was not to be enough, however, as Steve was soon to discover. In his second season at Portsmouth, the then club manager Bobby Campbell, a scouser from football's school of hard attitude and who was to take Pompey up from the Fourth Division, appointed Alan Ball as youth team coach. The man with a World Cup winner's medal from the year Steve was born, who would succeed Campbell to take Pompey back to the First Division (before taking them back down), and whose coaching Steve admired, decided that the lad's attitude was wrong and released him six months before his apprenticeship was up.

It stunned Steve, even if he did believe he would go on to make it somewhere else. Perhaps it was the best thing that could have happened to him as he would absorb the experience and redouble his efforts to become a professional footballer.

Life was always going to be unusual for me, I guess, given the way it started out. I was born in a naval officers' hospital in Portsmouth. My mother's father, I believe, was a high-ranking officer, who obviously had some clout. I have since met my blood mother, but not yet a step-brother and -sister on the Isle of Wight.

At the age of six weeks I was adopted by Anne and Alan Claridge, who I have always thought of as my real parents. Being adopted has never been an issue for me. I never had a need to meet my blood parents or felt deprived or sorry for myself; in fact, quite the opposite. I consider myself fortunate to have had such parents when I could have been stuck in a home; parents who have supported me through thick and plenty of thin in both my life and my footballing career as strongly as any family ever could.

We lived at 233 Huntspond Road, Titchfield, near Pompey, near my father's market garden business at South Leigh Farm, named after the village in Oxfordshire where he was born. Later he built the family home at South Leigh, a lovely spot overlooking open fields with the Solent about a mile away at the end of our road. It has been my bolt hole many times during my career and I love the place so much that I had a house built next to theirs ready for my retirement from playing.

We were never really poor but I think it was a struggle for him sometimes. That was the nature of his business with the vagaries of the weather and the market dictating to us. He once employed twelve people but he found it difficult to make a really good living. The house had only two bedrooms, my sister Ruth, also adopted two years after me, getting the second one upstairs so I had to sleep in the living room. We had no heating and it was so cold in the winter that if the freezer was full I think they put stuff in the living room with me to keep it fresh. I suppose *Monty Python*'s four Yorkshiremen talking about how times were 'ard when I were a lad would have considered it good for the constitution.

My primary school just across Huntspond Road was St John's and it witnessed the only time I have ever hit a woman. My mum told me that little Jessica, or whatever her name was, thought I was a lovely little boy, which seemed to incense me as I couldn't be doing with all those girlie things. I promptly went into school the next day and gave her a punch on the nose, apparently. It was probably a preview of my reaction to some of the football managers I have had down the years.

From there it was on to the local comprehensive, Brookfield. Now I have never been one of life's intellectuals but school really held no interest for me. I sensed pretty quickly that I would have to exist on my wits rather than any academic ability, that I would be best in situations where my judgement would be required. Though I say it myself, I became adept at thinking on my feet. I had to. I was hopeless at all the lessons. I don't think I was thick, just never able to see how triangles and trigonometry would help me score goals. Mind you, a few of the training routines I have been subjected to down the years bear a similarity and required a qualification in advanced algebra. I spent as much time out of school as I could, often cutting lessons to go with some people I knew in the market gardening business selling strawberries out of a minibus up the M27 for a bit of pocket money.

They say school days are the best of your life but for me they were the worst. Thank God for whoever invented sport. Without it I would have had nothing in life. It was all I really wanted to do at school. Sometimes when a teacher was calling the register in the morning and saw me there, he would say, sarcastically: 'Ah, must be PE today, Mr Claridge.' Football was obviously my first love, and becoming a professional was all I ever wanted from quite an early age, but I turned my hands and feet to anything and everything, with some encouragement from the games teacher Barry Stairs.

Most sports seemed to come naturally to me and I think I knew even then it would have to be my niche in life. I had a real scare at the age of 12, though, when I was playing a football match one day. I'd had a little rush of heart palpitations before but never really thought about it. Then, during this game, my heart started to go crazy, and not because I was having any feelings for Jessica, or whoever. I felt like I had just finished a marathon, completely shattered, although I shouldn't have been tired. I just sat there in the middle of the pitch with the game going on around me. Everyone was yelling at me to get up but I couldn't. It wasn't painful, just draining. Eventually I dragged myself up and tried to carry on but this rush just kept returning.

It was a very worrying time. I went for tests and they strapped a heart monitor to me for twenty-four hours. I was put on ten or eleven different medications over the next six months, none of which really seemed to help. In the end they gave me a tablet, called Amiodarone, which I was supposed to take every day, but it knocked me out. In consultation with a doctor it was agreed that I would double the dosage and take it only before playing any sport. Thankfully it worked and has done ever since. More than twenty years later I still take a tablet before every match, which keeps the condition under control.

As a freckly-faced kid, I also played basketball for the South of England, cricket for Hampshire Colts and tennis to a reasonable standard, often on a Sunday dashing from a football match to county tennis sessions at Bournemouth in the evening. One summer, as a teenager, I won four tournaments on a local circuit and managed to reach the final of the prestigious Lee-on-Solent competition a few miles down the coast from my home, where I lost to Chris Bailey, who went on to play at Wimbledon and for Great Britain in the Davis Cup before injury ended his career and he took up commentating for television. A few weeks later I reached the final of a tournament in Southampton, where I had to play the Hampshire no. 1 at that level, whose name I withhold to spare a lot of embarrassment for him and me. He never went on to be anything in the sport.

I had never really had any formal tennis coaching; I just picked the game up. But my rival was obviously being groomed for stardom. I turned up with my Woolworth's racket and dodgy trainers, while he came armed with seven rackets, a towel and an ice bucket for his drinks. With his two-handed backhand, he looked like Jimmy Connors. With his temper, he behaved more like John McEnroe, throwing tantrums all through the match, obviously not pleased to be losing to a scruffy oik like me. I duly won 6–1, 6–0 and at the end he threw his racket at me. I threw back a punch, which seemed to have the desired effect of shutting him up, though I don't think it was the done thing in such circles. Tennis and cricket in the county at that time were not really open to the masses and seemed more associated with coming from a money-oriented background. I didn't really like the people. They were different types from those in my first love of football, who were much more down to earth and among whom I felt more at home.

It was strange, but the only thing I wasn't picked for at any regional level was football, which really annoyed me. I made the school team all right, playing in midfield or up front – sometimes both. Being, I think, the team's best player, I went wherever the ball was. I enjoyed some memorable days, like a semi-final of the Hampshire Cup against a school in Basingstoke, in which I scored five second-half goals after we had been losing 4–0 at half-time. Mum and Dad had also started a club side in which I played, the Lock's Heath Sovereigns (named after a strawberry), providing all the kit, and we cleaned up locally. I think they thought that because it was the only thing I was really good at, and was the sport I enjoyed most, they should play to my strengths.

Then, at the age of 14, I was spotted by the Portsmouth scout Dave Hurst, who also later found Darren Anderton, and I was taken on by the club on associate schoolboy forms. This is a side of the game that the public rarely sees. In all, there were about 240 of us in the district on such forms, with the club, like every other, desperate not to let any talent slip through the net and keeping their options open on all of us. Every lad is starry-eyed, convinced he is going to make it. But over the next two years, I saw so many come and go. On training nights it was chaos, with boys changing in every nook or cranny available. If you didn't get there early – and quite often I didn't – and couldn't get into the two dressing rooms, it meant the corridor or by the side of the pitch. Naturally enough, there was a lot of 'showboating' going on with lads playing for themselves rather than a team in a desperate attempt to get themselves noticed. I don't suppose I was much different.

Dave used to drive me home afterwards and I recall one night unwittingly causing an accident, which was to be the first of quite a few I have been involved in. He pulled the car to a stop at a roundabout to let me out and as I opened the door on the passenger's side, a cyclist came up on the inside and crashed into it. Dave shouted to me too late. The poor lad dislocated his shoulder. I knew how he felt. Soon after that, I rode into the back of a car myself, went through the rear window and sustained some nasty cuts and bruises.

To me, Portsmouth's Fratton Park was Mecca at that time; it seemed a huge stadium to an impressionable young lad. Occasionally I would stand on the terraces singing the Pompey chimes with everyone else – 'Play up, Pompey; Pompey play up' – and I can remember watching a teenaged Mark Hateley's career, one which would take him to AC Milan and Monaco, take shape. I would watch all the strikers closely and even though Pompey were in the old Third Division, climbing back from a period of being down on their luck, players like Alan Biley and Alan Rogers seemed like gods.

What I remember most, though, is some of the apprentices taking the mickey out of my underpants on those times when we associated schoolboys were exposed to them and the cockiness that comes with being a junior pro. It was my first introduction to life among the lads and I suppose a necessary preparation for all the joshing you would be in for. I was desperately embarrassed when my boyish Y-fronts became the focus of laughing attention. I went straight home and asked my mum to go out and get me a supply of whatever they were wearing, jockeys or boxer shorts.

When I left school at the age of 16 I was one of four lucky lads offered an apprenticeship on £20 a week, going up to £25 in my second year. It was a great feeling; it felt like I had really made it, though I have to be honest and say that I thought I deserved it. I knew I was good enough and I don't think I had ever considered the alternative to not being taken on. It seemed a natural progression to me. Nowadays, clubs take on something like fifteen boys on Youth Training Schemes because the government pays half the cost and it represents cheap labour to get things done around the club. But you know that only one or two will be lucky enough to become full professionals. In those days, because it really meant something to be chosen from all those kids, you really felt you were on the verge of a glittering career.

That said, we were still cheap labour. It was a demanding life but a rewarding one, my introduction being pre-season training at *HMS Mercury*. It was hard but the camaraderie was great fun and the facilities were excellent. It was just fitness work, running from early in the morning, finishing with swimming in the afternoon, but, as with everything to do with the services, it was well-organized and structured, and at the end of the week you felt a million dollars. It was also quite an awe-inspiring experience to be around senior professionals like Billy Rafferty, a really good striker who scored pots of goals at that level.

I also had to undergo the initiation ceremony conducted by the second-year apprentices. I was stripped naked and boot polish was applied to every part of my anatomy – and I mean every part – with a shoe brush, though a fairly soft one, fortunately. I then had string tied round my wrists and genitals and was led out blindfolded on to a training pitch and tied to a goalpost. Then a sprinkler hose was turned on and I was left for what seemed like hours. I think the idea was that the water would wash off the polish but it didn't work out like that. It took hours of bathing and for days I was finding areas still black. Ah, the male bonding process, the rites of passage . . .

At first the club put me in digs but, lonely and homesick, I really couldn't stand it and after six weeks I moved back home. It meant leaving home at 7.30 a.m. for an 8.30 start, with two hours of jobs to do before training began. These days at a professional club, you turn up for training and the balls are dirty; then, I had to clean thirty-four of them every day. Each apprentice also had six sets of players' kit for which he was responsible, making sure it was all laid out properly each day. And we had to keep the boot room and all its contents clean.

We would train with the first team and reserves at Eastney Common, a bleak and windswept place, from 10.30 to midday, just making up the numbers, really. It was a bad system, I thought, because in that first season there was no full-time coach to look after us or teach us anything and we were left to our own devices too often, just kicking a ball around. The club manager was Bobby Campbell, who went on to manage Chelsea. He was a fearsome sort of bloke, though we didn't have much contact with him, thankfully. You never mucked around in his eyeline.

After lunch it was back to Fratton Park where we did more

fitness work, like running around the pitch, or swept the terraces till home time. If the weather was nice, we would go down to the beach at 4.30. Pompey used to employ four old blokes to sweep the terraces but one by one they died and the club never replaced them. We apprentices did our best to keep them alive so we wouldn't have to do it ourselves, feeding them vitamins and such like, but we just couldn't halt the march of time. Then, after night matches, we could be replacing divots until 11 p.m. In the summer we would also have to paint the ground before we were allowed our three weeks' holiday a year.

I was supplementing my meagre income working for my dad. One summer, I thought it might make a nice little earner to go up to the tennis at Wimbledon and try to sell strawberries outside the grounds. I remember getting up at 4 a.m. to pick them with a couple of mates and we set off up the A3 with fifty-two trays in the back of a van. Could we find a pitch, though? We just kept getting moved on, either by the police or other traders. We tried to sell them to the queues but they weren't really interested at breakfast time. It didn't help that another strawberry seller was following us around telling everyone that ours were full of maggots. We ended up driving to a council estate near Guildford that one of the lads knew and selling them there for a smaller profit than we had hoped for from the toffs.

Jason Pearce, Mark Davis and Darren Brown, the other three taken on with me at Pompey, were a motley bunch, good lads one and all. Jason Pearce was the most unusual, at least to the rest of us who were one-dimensional in our desire to be footballers, as he didn't want it as badly. He was heavily into the music scene and even wrote songs, entertaining us when we were sweeping the terraces, using his broom as a guitar. He seemed to know all about bands who were going to be the next big thing, like

Men At Work and the Psychedelic Furs, who meant nothing to us. Later he went to America to try to get into the music business. Now he is a policeman in Portsmouth. The others were Mark Davis and Darren Brown, neither of whom made it as a pro. The last I heard of Mark he was kicking around the non-league scene with Gosport Borough, while Darren became a postman. If it is difficult getting into the game in the first place, that illustrates how hard it is beyond that to get a full contract. And we were supposed to be the cream of the crop in the whole district.

I was in awe of the second-year apprentices, especially Kevin Ball, who went on to have a good career with Pompey and Birmingham City, among others. He was a frightening bugger. Some days he would force you into the boot room, which measured about 12 feet by 10 feet, with a ball for a game of one-touch football, which sounds more sophisticated than it was. He called it 'murder ball'. It consisted of Kevin wellying the ball as hard as he could. You were just trying to avoid injury but to have refused would have meant a worse injury – the blow to your pride and a weakening of your standing with your fellow young pros.

Of the senior players, I suppose I remember best Neil Webb and Ernie Howe. Neil was the bright young thing of the club at that time and it came as no surprise that he went on to star in midfield for Nottingham Forest, Manchester United and England. He had a lot of class on the ball. Then there was Ernie, a tall central defender who came from Fulham, where they used to sing: 'Six feet two, eyes of blue, Ernie Howe is after you.' He was never dirty, though; he was a lovely bloke and was always encouraging me. As an apprentice it was quite rare to get help from a senior pro.

On Saturday mornings we would play in the youth team in

the South-East Counties League, which was a good competition as you got to play against some of the top clubs, like Tottenham and Arsenal. Since there were only nine apprentices on the staff in all, we also had to field a few triallists, however. The coach in that first season of mine was Ray Hiron, who never really fancied me and I never thought much of him. Worse players than me, who never went on to anything, were getting games ahead of me, which really rankled. I never felt I learned much from him.

Ray, a lanky man, had been a successful centre-forward himself down the years for Pompey, one of their legends even, and perhaps he judged me harshly because he had played in the same position as me. By now I had settled on being a striker (with other good players around you – unlike at school – you don't have to do it all yourself). Perhaps Ray was also more on my case than other players because he saw some talent that maybe I wasn't fully developing, but at that age such thoughts do not really occur to you; you're not mature enough to think it through like that. You just perceive what you think is an injustice and dwell on it. In many ways, you are just living on your instincts. I also found it hard being told what to do. It wasn't that I was wild, just immature. Given all the clubs I have played for since, and all the managers I have dealt with, I think nowadays I could play for anybody. My experience has been so wide that I have probably seen it all and nothing really fazes me any more.

My timekeeping, I have to admit, was also erratic as a trainee, and I was often late in the mornings, not only because of getting the train in from Titchfield and the vagaries of British Rail. One time I can remember cycling to the station full pelt, coming off the bike and grazing my face very badly. Portsmouth were also getting fed up with me not attending the day-release course at Highbury College. It was quite right, looking back, as they were trying to look after other sides of your education beyond football – in case you didn't make it – but I had never enjoyed school and I wasn't about to go back.

In my second season Alan Ball was brought in by Bobby Campbell to be youth coach. He was great, Bally, and I did view him as something of a hero, as all young players interested in English football did. I thought he laid on inspired training sessions. He also saw things brilliantly, had such great passion and vision for the game, picking out things that were beyond almost all of us. The problem was he was so much better than everyone else and I think he got frustrated that nobody else could do the things he could with the ball. It must have been hard being a World Cup winner with England, playing for the best clubs in the country, like Everton and Arsenal, then having to come and sort out the likes of us.

Anyway, in the autumn of that second season he called me in. The timekeeping and the college absences were mentioned, but basically he didn't think I was going to make it. He said they were going to pay me off with £500, which seemed quite generous. I suppose I should have been devastated but I wasn't. I had always dreamed of playing for my local club, as most kids do – or at least did, in those days before every street in the country had a Manchester United shirt walking down it. But then, I have always had confidence in myself and my ability and thought that if I didn't make it there, I'd make it somewhere else.

Looking back, I was too cocky, too sure I was going to make it. I had this dream and I didn't think anything was going to disturb it. Being nice to myself, I suppose it was single-mindedness, but with that has to go hard work as well. Ability alone is never enough. I have lost count of the number of people who have come up to me over the years and said: 'I could have been a professional footballer. I was brilliant as a kid.' I just look at them and shake my head. If they could have been, they would have been. They may have had the skill, but they would have fallen down on something else. Perhaps they didn't want it enough, weren't tough enough physically or mentally to make it. They didn't have the dedication, drive or will to give up everything else. It takes a lot.

Then, I didn't really know all that – which was probably fortunate, or I might have given up at the first setback. Maybe it was a blessing that I wasn't too devastated. It certainly wasn't a blessing, though, that I went straight to the club offices and asked them to give me cash rather than a cheque so that I could go to the bookie's. Back in Titchfield, I blew the lot in three or four hours, which just didn't seem as serious as it should have, either. Easy come, easy go.

A month or two later, I started to realize how serious it really was. I was chosen for the county Under-18 team, which mostly featured players who had not found clubs for themselves, for a match against Yorkshire. It was at Fratton Park and I really wanted to do well to make a few waves locally and perhaps show Portsmouth what they had let go. Instead, I broke my ankle badly and could not play again for nine months. Suddenly it hit me just how precarious a position this was. I had always wanted to be a professional footballer, had been single-minded in that, and now there was a real danger that it was not going to happen.

I drifted throughout those nine months, getting into a few scrapes. That autumn I remember going up to see England play East Germany at Wembley – Bryan Robson scoring the winner – and on the way back overheating my dad's Rover 3.6 at 77 miles per hour on the motorway and spinning into the central reservation. It took me a while to get out of the car with the doors having jammed, and as I scrambled to the embankment by the hard shoulder, the engine blew up. I managed to phone Dad, who was pretty angry, but came to collect me anyway.

A few months later I missed a T-junction and went straight into a hedge, mangling another of my dad's cars. Then there was the time, soon after, I was driving along a dual carriageway in Fareham, in the outside lane, when a Volvo in front of me went into the back of an ambulance that was turning right, and I skidded into the back of the Volvo. His car was fine; mine just crumpled and was written off. I hit my head on the windscreen, stumbled out of the car and fainted into some bushes from which a couple of kindly old ladies dragged me. Another time, on Dad's farm, I was driving his tractor and eating a melon, which I dropped. As I reached down to pick it up, the tractor veered off and I ended up trapped in the cab in a ditch. I suppose I was lucky it didn't fall on top of me.

All this continued later into my professional career. At Cambridge United after training one day, I was eating fish and chips in my car and had left the vehicle in gear, with my foot on the clutch. When my foot slipped, so did the car and I hit the back of another lorry. During that spell I also jumped a red light in a little old Volkswagen Golf I had and hit a brand-new Porsche, midships.

If that was my fault, an accident that happened while I was with Birmingham City certainly wasn't. One night big Ted McMinn, a fans' favourite on the right wing, and I decided to go to Nottingham races. Ted said he knew the way and I followed him in my brand-new Vauxhall Calibra. Nearing the course, we got stuck in a jam, Ted got impatient and decided to do a three-point turn and find an alternative route. As he executed this manoeuvre, a car came the other way, saw him too late, swerved into the space he vacated and hit the car that was behind him – mine. I was just watching all this, doing nothing in particular when, bang, I was in an accident without even moving. I have also been stopped for speeding several times. My dad said, once, that it was because I had long hair. So I shaved my head. I got stopped again, and he said it was because I had a shaved head. It was probably just because I was speeding, though.

It's not that I am a bad driver – just accident-prone – and one who does 50,000 miles a year these days. I have been in quite a few prangs, but I generally hold my hand up if I am in the wrong, unlike my mum. She will say she has never been in an accident but she has probably caused about forty, totally oblivious to the trail of destruction going on around her. She has been known to go round a roundabout the wrong way.

That winter, spring and summer out of the game was a depressing time. I went for some training as a bricklayer but was just not suited to it. Most of the lads would have built half a house by the time I had laid a few bricks. Having been told by Alan Ball that I was not going to make it as a footballer, I was now told by an instructor that I was never going to make a brickie. I was starting to feel rejected.

I filled in my time by watching football, being a keen follower as well as player in my late teens. A cousin used to support Leeds United and I would often travel up to watch them with him, our petrol for the journey secured by siphoning it from other cars, which I suppose made me well qualified for the team I did end up supporting. Once we were driving up the M27 and some maniac with a shotgun ran on to the carriageway ahead and pointed the shotgun at us as we swerved to avoid him, before stopping off to phone the police and tell them about the incident.

Bored with Leeds, one day I decided to support the next team to beat them. It was Liverpool and that spring of 1984, dressed in replica shirt, red and white scarf and brandishing a Union Jack, I embarked for Rome on a twenty-four-hour adventure to see them play Roma in the European Cup Final. It was a frightening experience.

With my cousin, I drove to Dover and caught a ferry for France, then a train to Rome. It was a long, boring journey and we reached Rome a few hours before the match. Trouble was brewing even as we made our way to the Olympic Stadium with Italians pulling at our scarves and trying to steal our flags. Afterwards, Liverpool having won on penalties with Bruce Grobbelaar's wobbling knees my abiding memory of the shoot-out, visiting fans were being herded out when bottles and stones started raining on us from over a wall. There were a lot of cuts and bruises.

Once in the street, Italians started running at us but quickly dispersed when we confronted them. They were just not up to it. At this point we were divided up into modes of transport back home: plane, train or coach. I got caught up in the coach crowd when an Italian came over trying to swop his Roma memorabilia for mine. When I declined, he turned nasty and shouted to his mates who arrived to sort me out, one of them brandishing a small knife. I hit the geezer with my flag pole then turned and ran, one of them kicking me in the stomach as I made my getaway. I clambered on to a ledge and through a hole in a fence, scared stiff as seven or eight of them chased me. Luckily I ran into a group of Liverpool fans who charged at this bunch and sent them on their way. When I eventually got to the railway station and found my cousin again, our train had been wrecked, Roma fans having hurled stones through all the windows. It had been the same with the coaches near the stadium. It was a cold and miserable trip back to Calais through the night.

I firmly believe that the seeds of the Heysel Stadium disaster, involving another Anglo-Italian confrontation a year later when thirty-nine Juventus fans were crushed to death in a charge by Liverpool fans, were sown that night in Rome.

Apart from that experience, watching football was fine, but I really wanted to be a participant rather than a spectator. I was determined to try and get back into the game, but it got worse. I kept fit by running again on the roads around my home but did too much, incurring a stress fracture on the injured ankle.

At the start of the next season, I went to my local club Fareham Town and asked for a game, but I was still not fully fit. I couldn't even get in their first team and ended up playing in the reserves. I remember going to play a match on the Isle of Wight against Ryde Sports reserves, with five people watching the game. I was getting lumps kicked out of me. And all for a few quid. I was so gloomy by now and having doubts about what my future held. At the age of 18, with another winter of discontent looming, I began to feel that life and football was passing me by. I had to wise up pretty quickly and do something.

3 Go South-West, Young Man

Trevor Parker's motives were not entirely unselfish, he admits. 'Whenever I have had boys who want to be pros, and I see they might have the chance, I do all I can to help them make it,' he says. 'I know I might be losing a player, but in the longer term they might just come back if they fail, knowing I had done my best for them. And I know then I have a real good one who will have benefited from being around a pro club.'

Steve Claridge was one such. Parker, for many years manager of the Wessex League club Bashley, was surprised to hear in that autumn of 1984 that Claridge had been struggling at Fareham Town, where he was non-contract and not doing well enough to earn one. He hoped he might eventually sign him for his own team but recommended him first to Harry Redknapp at nearby Bournemouth, then a decent but perennially hard-up Third Division club surviving on gates of around 3500 and who were always on the look-out for promising and cheap talent at Dean Court. 'I don't think Harry took me seriously,' says Parker. 'But everyone in the locality knew Steve could play. It was just a question of bringing it out of him.'

Redknapp, a sharp and smart cockney who would go on to find the perfect niche as West Ham United's manager, having been one of its favourite sons as a player, invited Claridge for a week's trial but made up his mind to sign him within an hour of seeing the striker go through his repertoire. Stuart Morgan, then Redknapp's assistant, recalls the day vividly: 'We were doing a lot of one-on-one and two-on-two routines and Steve just tore the arse off everybody,' he says. 'We put him up against big Roger Brown, who was a good centre-half and went on to Norwich City, and he turned him inside out. Harry and I just looked at each other as if to say, We've got something on our hands here. He really had the bit between his teeth that day.' Indeed, such was Claridge's desperation to get back into the professional game that he trained as he had never before and perhaps never has since.

Progress was slow for him after he had officially signed on November 30, however, and though he recorded a personally memorable landmark by scoring his first league goal, against Newport County later that season, Steve could rarely get past the regular strikers Billy Rafferty and Colin Russell into the Cherries' first team. Mostly he was back on the non-league beat, playing in front of a few hundred for the reserves.

'After that first day when we thought he could be a world-beater if he could do these things to experienced players, we began to realize that Stevie didn't know quite enough yet,' says Morgan. 'We also thought there might be a glitch in his make-up. He was not a naughty boy, he just seemed to have a few problems. At pre-season training the next summer, for example, it was baking hot, about ninety-five degrees, and Steve turned up with a woolly bobble hat on. Harry told him not to be so daft and take it off, but he wouldn't. Steve said he had a problem. So Harry took him into his office and when Steve took the hat off he had only half a head of hair. He'd cut great chunks of it off.'

Steve these days explains the incident by saying that he had got paint in his hair and had tried to remove the offending section but miscalculated in front of the mirror. At first amused, Redknapp and Morgan became alarmed when it went on. 'It took me a year to get my hair right,' says Steve. But the management pair still had their doubts. 'We were playing an away match and decided to room Sean O'Driscoll, as a seasoned, sensible pro, with Steve. Then Sean came rushing down to dinner to tell us Steve was at it again upstairs. Harry went up and saw these great clumps of hair in the sink with Steve's hair now looking ridiculous and told him he had better shave the lot off. Steve did, and I think that may have been how his shaved-head look started. It seems funny but we thought it was just strange. We even rang Steve's mum and dad up about it and they said he did it at home as well.'

Morgan moved on to manage Torquay United in 1985 and early in the 85–86 season received a phone call from Brian Godfrey, the manager of Alliance Premier League side Weymouth – whom Morgan had previously managed with some success before going to Bournemouth (leading them to a famous FA Cup victory at Cardiff when they won 3–2 from being 2–0 down at half-time). Godfrey, the former Aston Villa inside-forward, said that he may be able to get one Steve Claridge for £10,000 and was enquiring about his abilities. 'I told him to get him if he could. He would murder that league,' says Morgan.

'I think Harry felt he had gone as far with Steve as he could,' Morgan adds. 'He was not in the team at that time and Harry may have been thinking he was not up to it.' Indeed, Redknapp, who had given Claridge only seven first-team starts in a year, may have tired of what he felt was strange behaviour. Morgan comments: 'Steve always had one sock up, one sock down and would take the field with one Adidas boot on and one Puma, one with screw-in studs, one with moulded. At ten to three he would still be in a corner of the dressing room by himself deciding which boots to wear from an old bag of them he carried. From a professional's point of view, it was hard to put up with.'

After being on loan at Weymouth for a couple of months, Claridge finally agreed to drop down to the semi-professional game, though his money went up, with the club then on the crest of a wave. It also gave him the opportunity to supplement his income by selling his father's fruit and veg from a stall near his home in Portsmouth and by taking on a series of gardening jobs, all of which resulted in the sack.

He was an immediate hit with fans of a club with a good tradition in non-league circles, even if the town is better known as the seaside resort patronized by King George III, whose statue stands near the esplanade. Weymouth had given Frank O'Farrell, who would go on to manage Manchester United, his start – him leading them in return to two Southern League Championships in the mid-sixties – and had produced a few Premiership stars of the eighties and nineties.

In 1987, Brian Godfrey, who had had two mid-table seasons after taking Weymouth to 5th in his first year, was replaced as manager by Stuart Morgan, back for a second spell and delighted to inherit Claridge among his charges. 'I built the team around him in many ways,' he says. 'He still had all his old character traits – often we would jog out on to the pitch at five to three without him while he was still deciding on what boots to wear – though I wasn't really aware at the time of his gambling. It didn't bother me anyway. He was doing the business for me so I just let him get on with it. He was an awkward bastard for any defender. You could knock the ball up to him at any height, angle or pace and he would somehow bring it down. Defences at that level just didn't know how to handle him. Mind you, I'm not sure Stevie knew how he did things half the time.'

Morgan immediately took Weymouth to the top of the Alliance, now named the GM Vauxhall Conference, with promotion to the Football League a distinct possibility in the autumn of Claridge's third season with the club. 'The turning point came for us around the New Year,' the manager recalls. 'We were second to Barnet, who we were playing over the holiday period and would have gone top again if we had won. We were expecting a crowd of around 5000 people but the pitch was flooded and the game was called off. It stayed flooded for a long while after, we didn't play a home game for three months and we lost all our momentum.'

By the end of the season, aged 22, Steve was pining for league football again, which he realized he was not now going to get with Weymouth. The belief that he had to leave the club was confirmed when, to his disgust, Morgan left him out of the Final of a good non-league competition, the GMAC Cup, against Horwich Railwayman's Institute. They lost. I was at Weymouth's ground early in the 1996–97 season and ventured into the programme shop, where I found an old copy of that night's match magazine. As I proffered my money at the till, the seller recalled the episode: 'Ah, Horwich, blimey. That was the night Morgan left out Claridge.'

Morgan explains: 'He hadn't really been doing it for us for a few games, had gone without a goal for a while, and I talked long and hard with my staff about it. I thought I would keep him on the bench and bring him on to do the business if we needed him, hoping he would be hungry. Looking back, he was up for the Cup anyway and perhaps I should have left him in. I know it upset him a lot. It was the only time there was ever a problem between us, and what made it worse was that it was the last game of the season.'

Claridge's last game for Weymouth, too. 'He was on my retained list,' says Morgan, 'but he didn't get the letter from the club informing him in time and he became a free agent. I took a hell of a lot of blame for it, although it wasn't my fault the administration didn't send it out in time. Steve was experienced and clever enough to exploit the loophole and knew he could better himself. It was a massive blow to the club, losing him.'

That summer several clubs contacted Steve, including Barnet, then managed by Barry Fry, Wycombe Wanderers, Bristol City and Oxford United. And Basingstoke Town, now managed by Trevor Parker.

'I had always admired him from afar,' Parker says. 'From Bashley I went on to Fareham, who reached the semi-finals of the FA Trophy under me. We were paired with Weymouth in the FA Cup and lost in a replay to them, with Steve outstanding. When I moved on to Basingstoke and heard Weymouth had messed up his contract, I left several messages with his parents asking him to phone me. Finally he did one night, at about half-past twelve. I told him I would like to talk about him signing for me and he said, "Fine." I asked him where and when and he said Rownhams in an hour. So we met at this service station on the M27 at one-thirty in the morning. I think he turned up in some battered old car with a coat hanger for an aerial, different coloured wings on it and stretched elastic holding it all together.'

Though he was still determined to get back into the league,

Claridge agreed to play a few pre-season games for Basingstoke to keep himself fit and perhaps get himself noticed. 'It would have been a travesty if he had stayed, I have to admit,' says Parker. 'He was an exceptional talent. I knew it was only right to give him every encouragement to get back into the league. As luck would have it, my assistant Roger Baddeley had just come back from holiday in Spain where he met Steve Coppell, the Crystal Palace manager. Roger rang him up and recommended Stevie.'

Soon Claridge had signed for Palace. 'Steve Coppell rang me and Roger up a while after, saying he owed us and would we like a day's golf,' says Parker. 'On the appointed day, his secretary rang and said he couldn't make it but he would be in touch again. In fact when you rang, I thought that was him.'

They say the darkest hour is just before the dawn. By now I realized I was going to have to apply myself more if I really was going to become a pro. Trevor Parker at Basingstoke, who had always seemed to like and rate me, reckoning I should make a pro, rang me to ask how things were going at Fareham, where I was non-contract, having heard on the non-league grapevine that I was struggling. Not too well, I told him, and a few days later a call came from Harry Redknapp, the Bournemouth manager, inviting me for a trial after Trevor had kindly recommended me to him.

It is strange how people's paths cross time and again in football, which adds to the impression that though it seems like a mega industry at times, it is really just a big village. Full of anticipation and gratitude I turned up for my first morning at Bournemouth, to be paired up with John Beck in a one-on-one drill. Now John was a talented, passing midfield player at that time, and my first impressions were favourable, as this senior figure danced past me. But what is it about managers that makes them change so much from their playing days? George Graham, nicknamed Stroller as a player but a strict disciplinarian as a manager, apparently, is a case in point. Though there was no sign of the long-ball philosophy that would later terrorize the Football League, even then Becky was a fitness fanatic. We used to see him still running the streets when we were driving home from training. It caught up with him, though. Dodgy knees finished his career prematurely.

After the week's trial Harry offered me a contract at £50 a week. Delightedly, I snatched the pen out of his hand to sign it before he changed his mind. The money hardly mattered. With £30 going to my mum for bed and board and the other £20 in petrol, I was not going to get rich but I was just incredibly relieved to be back in football, and this was as good a deal as I could realistically have expected. Bournemouth were a good club at that time, though strapped for cash.

It went well enough. I knew I was not immediately going to make the first team and I contented myself with the reserves, playing in such places as Shepton Mallet and Glastonbury. Every day I would drive up for training from Pompey with my fellow striker Billy Rafferty, by now transferred from my home-town club, and the goalkeeper Ian Leigh, picking them up at the Concorde Club in Eastleigh in an old Morris Minor I had at the time. I remember one day carving a bloke up at a roundabout, unaware of having done so. As I pulled up alongside Billy, the bloke, who had followed me into the Concorde car park, got out of his car and set about punching me. Billy had to pull him off.

It was strange teaming up with Billy, who had been a first-teamer when I was an apprentice at Portsmouth. He wasn't exactly a hero of mine, but it did feel good to be alongside a player whose boots I'd cleaned a year earlier. But Billy was then on the way down. I remember seeing him a few years later and he told me about a club in Portugal he went on to where they had a superstar who couldn't last ninety minutes, so they would play him for seventy and bring on Billy for the last twenty. Nice work if you can get it.

I was pretty quick in those days, having done a lot of district athletics at school, and Harry Redknapp also knew I liked a bet. Next to the club's Dean Court ground there used to be a bumpy, grassy track and Harry used to make us do a 600 metres run round it during training. The club record was 1 minute 30 seconds but one day I did it in 1:24. Harry didn't believe me and bet me £100 I couldn't do it again. Naturally I took him on and all the club staff – players, youth team and management – came out to watch. My first mistake was trusting Harry with the stopwatch. My second was trusting him.

He said he would help me out by providing one of the kids as a pacemaker. The lad's job was in fact to try to slow me up. When I did get past him, I found that Harry had put out some signs as obstacles on the top bend, claiming he hadn't been able to move them, forcing me to go wide. At the finish, Harry told me I had missed the record by a second. When I complained loudly about all the obstructions, he finally wavered and agreed to let me off the £100. Only later did one of the other players tell me I had broken the record. And there was me grateful to Harry, which I suppose I was, just for being at the club.

I liked Harry and thought he was pretty good as a coach, though I was always wary after that episode. That first season I wasn't really ready to play in the first team but after getting a few games in a row towards the end of it, I thought I might have a chance of starting the next season. Harry thought otherwise. It was a typical situation with a manager, really. If you're in the team, you like him; if you're not, you don't. And that's how it should be. You shouldn't be happy with him if you're not playing.

Harry once tried to swap an old Datsun he had – he also wanted a cash adjustment – with my nice little Morris, which was a good runabout though I could never get it above 42 miles per hour. I later found out the Datsun had been abandoned in the club car park and about a week after I refused to be railroaded by him he scrapped it. Still, I will be forever in his debt. Someone told me a few years later that he had considered buying me when he became manager of West Ham but that he thought I was a bit off the wall. I suppose that day at the track sprang back to his mind, with me likely to claim back my hundred quid. Then again, perhaps I should have taken the Datsun off his hands.

I played only seven games for Bournemouth but I did score my first league goal. It was against Newport County. The ball came in from the right, Billy Rafferty's header was saved by the goalkeeper and I had a simple job to nod home the rebound. Not very memorable to many but it was to me. I was on my way.

Or at least I thought so. With Billy and Colin Russell as the first-choice strikers, I spent most of that season on the fringe of the first team – but then the way the club stood the groundsman was on the fringe of the first team. So, too, was this little 12-year-old who used to practise with us at the end of training. Name of Jamie Redknapp. He looked the part even then – good control, good passer – and has developed along the same lines with Liverpool and England. I only really got my chance late on because Colin had some sort of bust-up with Harry. I was playing mostly for the reserves at places like Taunton and Weston-super-

Mare in front of a few hundred and by now, the spring, not enjoying it very much. The beginning of my second season, when I wasn't near the team for the first few games, seemed to promise only more of the same.

Then one night we played at Weymouth, who were then in the Alliance Premier League, and they seemed to like me there. After the match, a few officials from the club came up to me and said they thought I had played well. A few days later the Weymouth manager Brian Godfrey, once a midfield player with Aston Villa, asked to take me on a month's loan, which Harry agreed to. I enjoyed it, things went well and soon they bid £10,000 for me. Bournemouth thought it was too good an offer to refuse and suddenly I was a non-league footballer twenty-five miles further west along the South Coast.

The prospect of being a part-timer did bother me but the money was good – £90 a week going up to £110, with a £2000 signing-on fee over the two and a half years of my contract. In many ways, Weymouth were a wealthier club than Bournemouth, and with gates of around 1000 and a new ground being built, they really did seem to be on the way up. The goalkeeper Len Bond's signing-on fee was £9000 over three years, which seemed like an absolute fortune at the time, and some of the lads were on £200 a week, more than full-time players were getting in the lower divisions. Anyway, I have always lived by the saying that you should go where you are wanted, rather than stay at a club who are uncertain about you, and I was being offered first-team football instead of the reserves.

It was a good move for me, a good-quality league to go in. I could have drifted around reserve team football for a long time. I was still a teenager and sometimes kids don't get taken seriously at lower-division clubs. The longer you stay, the more you get taken for granted. Players get signed and are put on more money and all you do is get poorer and go down in the pecking order. The only way to get a pay rise or more regular football is to move on, which is what some fans don't realize when they criticize players for lack of loyalty. Ten years holding on hoping for a testimonial – that's if the club want to keep you – is a long time, a whole career.

The standard at Weymouth wasn't a great deal different from the Third Division with the Alliance quickly showing itself to me as a tough, competitive league. The way of life was a change, though. I had been used to training every day but here it was Tuesday and Thursday nights, driving over from Portsmouth in my dad's old Granada and picking up a team-mate, Mick Doherty, a really prolific striker at that level, at Basingstoke. Mind you, I had never really enjoyed training, so it suited me well.

It also meant I had time on my hands and could supplement my income with other work. On Saturday mornings before games and on Sundays I would sell my dad's fruit and veg from a stall outside somebody's garden. I built it up to include Fridays, then started another one, selling strawberries in front of a field on a main road. I started off earning about £30 a week and increased the takings to £60.

A friend who had been at a horticultural college near Winchester told me he was making good money at gardening, so I thought I would give that a go as well. He knew what he was doing, though. I didn't. I got the sack from one job for digging up a woman's flower bed and fired from another, at a really big house, for dossing around when the owners were on holiday. When I thought the neighbours, who had obviously been asked to keep an eye on things, were watching, I would do a bit. When I thought they weren't, I would be walking the dog, swimming in the pool or watching Channel 4 racing. Unfortunately the neighbours were watching pretty much all the time, and once caught me having a swim. I was also sacked from one place after being found skiving behind a big tree. At another house, they asked me to build a roof for their conservatory but I fell through it. After the third experience, I felt I was being told that this may not be for me. Still, it meant I had more time for my stalls.

My first season at Weymouth was a good one as we finished 5th. I remember a great day against Dartford in the February, scoring 4 goals as part of my total of 18 for the season. Instead of building on it, though, we became a middle-of-the-table team the next season and at the end of it Brian Godfrey was replaced as manager by Stuart Morgan, who I knew from Bournemouth. He was all right, a decent bloke, but a bit dour and I always wondered whether we were going to win anything under him. A fair-to-middling manager, I would say. Unfortunately our really good team of my first year, when I enjoyed myself up front in a roving role alongside Mick Doherty, who was small, quick and sharp, and Tony Agana, who had some silky skills, was beginning to break up. Tony had gone to Watford and Tommy Jones, a classy midfielder, to Aberdeen.

Memories of some strange incidents abound, one at Barrow, the most bizarre of my career. I was minding my own business on the halfway line, as Barrow got a corner, when out of the corner of my eye I noticed the player marking me, a massive geezer, taking a run-up. A bit far out to attack the corner, I thought. As it turned out he was coming for me; one set of studs landed in my ribs, the other on my shoulder. I can remember thinking, I'm not having that, so I got up, ran at him and jumped. The problem was I just bounced off him, so I thought the discretion of staying down was now the better part of valour. It was just weird, not least because nobody else on the pitch seemed to notice the incident. If it had happened in a televised match, with a camera capturing, for argument's sake, a Frenchman at Selhurst Park, all hell would have broken loose. The consolation was we won 2–1 and he got sent off later.

Then at Barnet we returned to the dressing room after a match to find we had all been robbed. I had personally lost £200 in cash. As soon as he heard, Barnet's chairman Stan Flashman, a controversial character, who used to sack Barry Fry as manager every other week, came in and sympathized with us. He also did the decent thing by reimbursing us all. When I saw the size of the wad he got out of his pocket, I regretted having told him it was only two hundred that I'd lost.

It was during that season that I thought I was going to die on the pitch. It was at Altrincham and I was chasing a ball over the top when their goalkeeper Jeff Wealands, who had been with Manchester United, came steaming out and caught me under the heart with his knee. I am sure it was an accident, because Jeff was a good guy, but it must have looked like that foul by Harald Schumacher of West Germany, against France in the World Cup in 1982, which almost killed Patrick Battiston. It certainly felt like it.

Not that anything rational was going through my mind as I lay there nursing two broken ribs and gasping for breath. I had never before been in such pain. I was panicking but luckily the physio Bob Lucas calmed me down. He was a lovely old fellow, Bob, a very old-fashioned sort who seemed too nice to be in football. He had been a servant to the club for over fifty years, having played for Weymouth in their biggest ever match, against Manchester United in the third round of the FA Cup in 1950, when they lost 4–0 at Maine Road (Old Trafford having been bombed during the War). I saw him about ten years later and asked him how the club was doing. 'We've got a good manager now and things are turning around. We'll be out of the Southern League Southern Division soon,' he said. Two weeks later the manager was sacked. What kept him so young, optimistic and involved I'll never know, but thank God people like him are still in the game and keeping it going.

As well as Bob, there were some real characters at the club. Richard Bourne was just about the biggest defender I have ever seen; Giant Haystacks with heading ability. Alongside him at first was the scouser John Carroll, who went on to manage Runcorn. 'John Carroll's' he would bellow from at least 50 yards away from a high ball, and would make it as a path parted. No one dared to get in the way or challenge. Later we had Paul Compton, who went on to manage Torquay United, and Shaun Teale, who was quality. He was really quick and strong, the best centre-back in the league, who deserved the career he had subsequently with Bournemouth, Aston Villa and Tranmere Rovers. Then there was Anniello Iannone, a real Weymouth character. Of Italian extraction, he had a window-cleaning business in the town, was as strong as a bull and had a stout heart. The fans loved him and he was a good servant to them, playing anywhere and everywhere.

There were also the Rogers brothers, Peter and Martyn, market traders in leatherwear around the Bristol area. Peter was the practical joker of the team and did me up like a kipper within my first couple of weeks at the club. 'I hear you like a bet,' he said to me one day, handing me a folded-up scrap of paper. 'Here's a tip I got from a Chinese restaurant owner in the town. Keep it to yourself and don't open it till you get home.' When I got home I duly did. The writing was Chinese. Years later I was playing for Cambridge United at Plymouth when this voice from the terraces started shouting about what a tosser I was. It was the sort of comment you get ten times a match but something made me look across. There was Peter laughing and waving at me, and all I could do was grin back at him and concede he might have a point. The last I heard, he had gone full-time as manager of Tiverton Town in the Western League.

For the start of my third season, we moved out of the old Recreation Ground to the newly constructed Wessex Stadium on an industrial estate near the Radipole Lake bird sanctuary. All of a sudden the club took off. Although he'd lost Agana and Jones, Stuart Morgan brought in some really good new players, like the goalkeeper Peter Guthrie, whom Terry Venables would sign for Tottenham for £100,000 later that season, Peter Conning from Rochdale in midfield and Shaun Teale. It has happened at Middlesbrough and this was the same thing on a smaller scale: new ground, money to spend. You don't realize how many people are out there just waiting for something to support, something to aim for. Most clubs don't. We won our first five games, including a great night against Lincoln City when 3600 people saw us beat the newly relegated side 3–0, and the whole town was buzzing. More than 6000 came to see the official opening against a strong Manchester United team, which featured Bryan Robson and the newly signed Brian McClair but in all honesty were only going through the motions, when we won 1–0. I think the gate money from the first four games exceeded the whole of the previous season.

We all thought we were going to be in the Football league in a year. Though we lost our sixth game 2–1 at Maidstone – rivals for the title and who eventually made it into the league before going bust – we were still top of the league come November and looking good. Then in mid-winter our pitch sprung a leak, with a natural spring being discovered in the middle of the pitch, and it really knackered us. They should have started bottling it at source to make money. Instead, it probably cost the club a fortune as the pitch became a swamp. Other faults emerged because the ground had been built in a hurry: the away end was covered, but not the home; it was supposed to be segregated but home supporters had to walk past the away end to get in; the turnstiles were too narrow for anyone except Kate Moss; and the grandstand had no sides, so when it was windy and rained, which seemed just about all the time, the seats got lashed.

For nearly three months we did not play a home game and the money began to run out, the club having spent big in the hope of promotion then witnessing gate receipts going down as we travelled all over the South-West to fulfil our fixtures. We had to play at Bournemouth, Dorchester's old sloping ground and Poole, which was just about the worst place in the world for football. The pitch seemed to be surrounded by a running track, then a dog track, then a speedway track, and the few spectators who braved this bleak spot needed binoculars to see the game.

I particularly remember one game against Wealdstone at Bournemouth. Bad blood had existed between the two clubs from the previous season when we'd had a kicking match at their place and Vinny Jones – yes, that Vinny Jones – and Len Bond had both been sent off for fighting. It was clear this night that the incident had not been forgotten. Although the most famous hod carrier in the country was by now with Wimbledon, this was probably the most frightening match I have ever played in. I have never before, or since, been part of a game where players were not worried about the ball in the least. Safety first, every man for himself was the order of the night. Within minutes players were elbowing and kicking each other. It was hardly the referee's fault, I suppose, but faced with such mayhem he failed to establish any control and had to abandon the match for everyone's well being. Had it gone on, he probably would've had to anyway, with both sides having fewer than seven men each on the field.

I don't think we won a game in those three months. In fact we once lost seven in a row after our start of five consecutive victories. Our away form deteriorated badly and our challenge just petered out as we finished 10th. In many ways it was the turning point for the club. We'd possessed all the credentials to go into the league, it seemed, but the stuffing was knocked out of the club and it was soon in decline, being relegated to the Southern League a few seasons later, where it remains. It is such a shame because Weymouth has the potential to be a good football town, with no real competition for 25 miles – Bournemouth being that distance; Southampton 60 miles – and a population of around 60,000. With some success they would get good crowds, probably good enough for the Third Division. Then, we were one of the best non-league sides in the country and scouts used to come looking for – and find – players there. Before our team broke up, with many finding league clubs, the tradition was established with Spurs signing Graham Roberts and Southampton Andy Townsend.

They were right to do so because these were big names in the small world that is non-league football, where everybody knows everybody else, unlike at the professional level where everybody has merely heard of everybody else. There you don't quite socialize as much as you do after games at the semi-pro level. I would never knock it because although it does not represent people's full-time livelihoods, everyone plays it hard and there are some big matches. Some small ones too, like playing in the Dorset Senior Cup – Weymouth's annual saving grace for a trophy – at Sturminster Newton. Their ground is perched on top of a steep hill that some of my old cars would never have got up. The manager would not accept my excuse that I suffered from vertigo, however.

As a consolation that year, we did reach the final of a decent non-league tournament, the equivalent of the Football League Cup, against a team called Horwich Railwayman's Institute, but for me that marked the end. I had all but decided that, at the age of 22, it was time to make a more determined effort to get back into the league. My contract was up. I had missed only two games for Weymouth that season, was their top scorer and the supporters' Player of the Year, but after a four-hour journey up to Horwich, near Bolton, Stuart Morgan told me he was dropping me to substitute. I had my first real clash with the manager but I wasn't so much angry as bemused by the decision. I didn't even bother to ask why, I was so shocked. I was particularly upset because I had brought up several friends and relatives for the game. I knew after that Final, which we lost on aggregate, that I couldn't work with Stuart for another season, though I still considered him a good enough bloke.

Weymouth had also messed up my contract. They were supposed to send me a letter three weeks before the end of the season to say that they were retaining my services. That way, they would have got a fee for me even if I did leave – I was planning to reject any new terms – and I think I might have been worth at least £50,000 on the market. Dave Bassett had signed Tony Agana for Watford for £100,000 the previous summer, after all. My letter never arrived, though, and in the second week of May I was a free agent trying to find a club again. I think it caused quite a controversy at Weymouth, with the oversight being heavily criticized by the fans. Some people said it was the beginning of the end for Stuart Morgan, who did not last much longer at the club.

It was an interesting summer. The first to contact me were Wycombe Wanderers, really ambitious, as their progress up the league has since shown, and they treated me like a king. The chief executive and the chairman wined and dined me at a posh hotel and made me a tempting offer: £280 a week, a signing-on fee of £2000, rent-free accommodation and a job as a van driver to supplement my income. A lot of league players would have been envious. By contrast, Barry Fry, then manager of Barnet, who had just finished runners-up to Lincoln in the Conference and were getting desperate for promotion, phoned and asked if I would meet him at Fleet service station on the M3. He was wearing shorts, brogues and not a lot else when I met him on a blazing hot day.

'How are you son, all right? I think you're a fucking great player, I do,' was the opening, and typical, remark of a character I would come to know well at Birmingham City six years later. 'What you agreed with Wycombe?' 'Nothing yet,' I said. 'What they offered you?' was his next question. When I told him, he came back with: 'Ah, yeah. But that's before tax, innit?' He was offering £230 after tax, £5000 as a signing-on fee and a £50 win bonus. 'That's money in your hand,' he said. 'And Stan will see you all right.'

Barry and Stan Flashman had been the talk of non-league football, their desire to get into the league so intense. The word was that there was always plenty of money there and the sight of Stan's wad that time we were robbed at Barnet had proved it to me. Legend had it at Weymouth that shortly before our last game of my second season, against Scarborough, a voice claiming to represent Barnet phoned my striking partner at Weymouth Mick Doherty offering a 'wheelbarrow of money' if we won up there. At the time, Barnet were second to Scarborough in the table and if they were to go up needed us, then a mid-table outfit, to beat them, while they themselves were winning at Dagenham the same day. As it happened, we went one up at Scarborough against a team nervous at being on the brink, and the thought of the barrow load of dosh did go through my head, but we eventually lost 2–1 in front of more than 7000 people and Barnet finished runners-up.

Both Barnet's and Wycombe's offers were appealing, and Weymouth even came up with funny money to outbid them, probably desperate to rectify their mistake in the eyes of their supporters, but I wanted to go full-time again. I had trials at Bristol City and Oxford United, both saying they wanted to sign me but weren't in a position to at that time. The Oxford manager Maurice Evans was away on holiday and his coaching staff wanted to wait for him to get back so that he could have a look at me.

I played a few games for Basingstoke Town that pre-season to keep my eye in, and their manager Trevor Parker and his assistant Roger Baddeley did me another good favour, telling another contact, Steve Coppell at Crystal Palace, that I was available on a free. When he contacted me and then offered me a deal it became no contest. Palace were offering £300 a week – I would probably be worse off than at Wycombe – but £5000 a season signing-on fee over two years, which was a bonus. The real bonus, though, was being back in the league and with a club in the old Second Division with real prospects of promotion to the First. I could hardly believe my luck.

Barry phoned me for a decision. 'We've all got no chance, then,' he said accurately when I told him that Palace had been in. 'Good luck. See you again some time,' he added, also accurately.

4 Shots in the Dark

Crystal Palace are one of those halfway houses of the English game. Their pretensions are towards being a big club at the top level. The reality is often a brief spell there and relegation, followed by several seasons trying to get back up. For clubs like them with crowds not quite big enough to challenge the big clubs of the Premiership but too big for the First Division to contain them for too long – a number that also includes Bolton Wanderers, Ipswich Town and Queen's Park Rangers – perhaps they ought to form a Division One and a Half.

When Steve Claridge signed for the south London club in the summer of 1988, Palace's yo-yo was on its upward movement. They had been languishing in the old Second Division for eight years but, after four years in charge, Steve Coppell was now building a promotion team and Claridge could feel himself privileged to be in on it, given his previous humble environment. It did not look especially promising at the start of the season as Palace won only one and drew the other five of their first six matches, but eventually they would finish 3rd and beat Blackburn Rovers over two legs of the play-off final, overturning a 3–1 deficit with a 3–0 home victory.

The two main reasons for the success were Ian Wright and Mark Bright, who between them scored 44 league goals and another 4 in the play-offs. Two more in the cup competitions made it a round 50. They were also the two main reasons why Steve Claridge would last no longer than three months at the club, though there were others, including a falling-out with the reserve team manager Alan Smith, who would go on to be Coppell's successor.

'I think he was in awe of everything and everybody to start with,' says Ian Evans, then Coppell's head coach and who went on to be assistant to Mick McCarthy, the Republic of Ireland manager. 'He kept himself to himself and seemed very quiet. His problem was that he was not willing to do anything out of the ordinary, and it was a battle to get him to do more than the minimum. He didn't seem to want to put in the time after training to learn more. He was usually late coming in for training and pretty sharp leaving. You would think to yourself, He was here a minute ago, where's he bloody gone now?

'His ability was always there and once he got into the swing, he was quite receptive. As a coach you take to players who are eager to learn, and I got to like him. Mick took a fancy to him as well, later, and when we were at Millwall together we thought about signing him. He would keep the ball well and always want it, but sometimes you felt you had to throw another ball on for the others to play with because he would want to keep it. Gradually he got to become more of a team player, though.'

Any strange behaviour he noticed?

'His kitbag, mainly. It was always full of old boots and mouldy sandwiches and didn't look as if it had been cleaned out for twelve months.

'I was always cracking the whip with him because I thought he could become a good player if only he would jolly himself along, but perhaps deep down he thought it was pointless because at that time Wright and Bright were a bit special. I've always followed his career since and been pleased at what he's achieved.'

To try to prevent the fines for lateness that were eating up most of his wages, Steve bought a flat in Wimbledon. He never got to move into it. Coppell received a bid of £14,000 for him from the Aldershot manager Len Walker and took it. Steve was happy to move on, signing on October 13, his philosophy being that he would rather play first-team football lower down in preference to stagnating in the reserves of a big club. He was also delighted to find a motley crew of kindred spirits at the battered old Recreation Ground.

It was a charming setting. From the High Street, turnstiles housed in huts with patterned glass of the sort you might see on seaside promenades led up to the ground through floral gardens, which precluded terracing at that end. There, the Shots, in their unusual red and blue, would be just a yard or two away from you, warming up. The idyll – for a football fan, that is – of such a view masked deep problems at a club that no one knew, even if they suspected, was in its death throes. Also behind the scenes was the lower divisions' version of Wimbledon's Crazy Gang.

'We were a bunch of misfits all slung together,' recalls the Scottish defender Ian Phillips. 'Everybody had a skeleton of some sort in the cupboard. It was probably why everybody got on so well.'

Steve would be with Aldershot for fifteen months and would

recall it as the happiest time of his career, even though the club was always in danger of going out of business (it would do in 1992) and he went a total of fifteen weeks without being paid. Added to that were the poor training facilities and Christmas bonuses from the club of rotten turkeys. The place was full of characters, though, with Claridge chief among them. He even sold fruit and veg from the back of his car to colleagues.

'People say there are no characters left in the game these days,' says Phillips. 'They've obviously never met Steve Claridge. I remember his first day at training. He turned up in a battered green Escort which should have gone for scrap. It was full of rubbish then, and always was afterwards. We thought it was a skip. Whatever he read or ate was thrown down in it. He used to leave it unlocked, in the hope that someone would steal it, I think.

'He also used to hide all his cash in it because there was some problem with his bank account and the club couldn't pay his wages (when we got them) into it. Yes, we knew all about his gambling. I think he bet most of his wages. A few of us would talk about the horses and his ears would prick up. He would always tell you about it as well. If he was chirpy, you knew he had won, if he was down you knew what that meant too. If it's in your blood, it's a hard thing to kick.'

Claridge's timekeeping, too, was a source of amusement within the club, though in theory there should have been less trouble than at Palace, the journey from Portsmouth to Aldershot less demanding than that to south London. 'He was always rushing for the team bus,' Phillips recalls. 'You would be just about to go and he would arrive with some excuse about being stuck in traffic or some accident happening just in front of him. We used to say, ''Well, why don't you leave a little bit earlier, then?'' Len Walker and his coach Ian Gillard would just say, "OK, then. No fine." I think he got away with it because he was such an easy-going guy. There never has been any malice about him.'

Then there was the boot bag.

'In all my twenty years as a professional I never saw anything like it,' says Phillips. 'He would sit there in the dressing room for ages just staring down into this bag wondering which pair to wear. Sometimes they wouldn't be a pair. A psychologist would have had a field day. We used to say, "For God's sake make a decision." He would go out for the warm-up then come back in and change them. I have even known him start a game with a pair then after five minutes tell the bench he wanted to go off and change them. He might do it two or three times during the afternoon. He would never tie his laces very tight either and he'd sometimes lose the boot during a match.'

Despite it all, Claridge enjoyed a good season with Aldershot, scoring nine league goals and being voted Player of the Year. 'We always thought he would make it,' says Phillips. 'As long as somebody took a chance with him. He was always a very workmanlike player but did things out of the ordinary. His touch around the box was unlike others in the Third and Fourth Divisions. He was also incredibly fit, which sounds strange given his lifestyle, but I think it was just natural fitness. I can remember there was one hamburger place open in Aldershot until midnight and sometimes coming back from a game up north it would be approaching that time when we got home. Steve would always be the first off the bus and would sprint the half a mile or so to it flat out, while the rest of us could hardly move.'

Phillips recalls the turkey incident. 'Fortunately I didn't cook mine because the stench was so bad. But I think some players did and only realized the problem when they came to carve it. It was embarrassing, but then a lot about Aldershot was. Not getting paid for so long was a big problem and I had to go to the bank for an overdraft. Some of the players got quite militant but we all agreed to play in the end. The players were just such a good crowd, and Stevie played his part in that.'

Claridge was eventually sold to Cambridge on February 8, 1990, for £75,000 so that Aldershot could pay some bills and stave off their demise. He returned to play against the Shots later that season and Phillips confirms the story that there was a competition to see which home player could kick him the hardest. 'Well, not kick him exactly. Just to see who gave him the hardest, strongest tackle. I don't think it was the tenner that Steve remembers, more like a pound, but yes, I was the one judged to have given him the best tackle. I never saw any money, mind.'

Phillips went on to take a job with British Gas and also travelled from his Colchester home to play part-time for Kettering Town, before becoming manager of Halstead Town in the Jewson Eastern League. 'I kept in touch with Steve,' he says. 'Once he rang me up to get him a new fireplace and a central-heating boiler at cost price. I wasn't supposed to do it because it was for staff only but I did because he is such a lovely lad. Then we had the problem of getting it over to Luton, where he was living. He sent his team-mate Michael Cheetham to take it to Cambridge, then he drove it home himself. I think he had to have one of the car doors open, driving all the way to Luton with it like that.

'I always have a laugh with him. In fact, when I was choosing my Fantasy League team I rang him up to tell him that I was putting him in mine. He asked me how much he was valued at, and when I said £3 million he said he wasn't worth it. I'm really pleased that he made it to the Premiership in the end. He probably was more eccentric in his early career, but when he knuckled down he went from strength to strength.'

The idea of playing for such a potentially massive club as Crystal Palace was so appealing that I didn't really think it through properly. It should have been the dream move but it became a nightmare. I was with the club for less than three months and I ended up owing them money in fines for being late. Also, Ian Wright and Mark Bright were at the club at the time, and seemed bigger than it, so I was never really likely to break through into the first team.

The main problem was that I had to commute up and down the A3 from my mum and dad's house to Palace's training ground in the Morden area of south London, not the easiest place in the world to get to at the best of times. To top it, roadworks from hell on the Guildford bypass had just started. The saving grace was that with me not getting much money after all the deductions, and also spending most of the day in a car, my opportunities to indulge my gambling were distinctly limited.

I encountered my first problem as early as my third day there. After training, we were all told to go to Selhurst Park for the traditional pre-season photo session and, not knowing the intricacies of the route between Morden and Selhurst, I asked another player, Adam Locke, if I could follow him. He roared off and I couldn't keep up. I didn't know where I was going. At what I thought was a mile away from the ground I asked for directions and ended up about three miles away. When I finally did arrive, the whole club was waiting for me. I was not Mr Popular and had not created a very good first impression.

Neither did I at one of my first matches for Palace reserves. It was against Portsmouth at Selhurst Park and I really wanted to do well against the club who had let me go. My dad came with me to help navigate through the streets of south London, which even Londoners have trouble with. Someone once told me they had never been to Selhurst the same way twice, so convoluted were the routes. Anyway, I turned up half an hour late, though well before kick-off, and the reserve team manager Alan Smith fined me what I thought was an astronomical amount: one week's wages. It really took the wind out of my sails.

I was dropped to substitute and was resigned to not getting on. I was desperate to play as there were people in the Pompey team from my time there. I sat on the bench gutted. But by half-time we were 3–1 down and Alan Smith turned to me and said I might as well get stripped and come on for the second half. We got it back to 3–2 and then I scored the best goal I can ever remember scoring, better than the one for Leicester City in the play-off final. The ball came over from the right wing at shoulder height and with a scissor-kick volley I sent it into the roof of the net for the equalizer. Eat your heart out

Mark Hughes. Late in the game, I followed it up with a tap-in to give Palace a 4–3 win. It just summed up football for me in ninety minutes; you can be down in the dumps one half, then walking on air the next. In the dressing room afterwards, Alan Smith was very grudging. All I can remember him saying was: 'It'll be coming out of your wages.'

I never really got on with Alan, who went on to become Palace's first team manager then took over at Wycombe Wanderers. I am afraid I had no respect for him, probably because I felt he had little for me. He was a non-league manager who stepped up to coach Palace's youth team, though I have to say I saw him being interviewed on television a few years ago when he had taken over Palace's first team, then involved in a relegation battle in the Premiership, and I thought he handled himself very well, with a lot of dignity. People tell me he's a very likeable man, but I can only say that my impression at that time was that he didn't know enough about the game.

He also committed what I think is a cardinal sin for any manager, by treating players differently in the same situation. We used to play some reserve games at Tooting and Mitcham's ground, which I personally found easier to get to than Selhurst, and one night John Salako turned up late at Selhurst – only a quarter of an hour before kick-off – claiming that he had been to Tooting, thinking that the match was there. Alan didn't say a word against him, accepting his excuse.

I was new to the club, unsure of my standing, and didn't feel able to say anything; I probably would do now, though. But I did think, This isn't right. I am travelling all this way and getting fined. He lives on the doorstep and isn't. Players are very sensitive about what they perceive to be preferential treatment. Any coach must treat players with equal firmness if he is to have their respect. Of course coaches always have their favourite players, but they must never let the squad know who they are, or they'll lose the rest if they single them out. It is supposed to be a team game, after all. It's probably a bit like having kids; you may like them differently for different reasons but you have to be even-handed with them. And players are big kids, I have to admit.

Alan had some complicated methods of training, using cones and dividing the pitch up into grids. 'Right, I don't want you going out of your grid,' he would shout at me. Well, I am not a Gary Lineker, not essentially a penalty-box player. I like to get the ball wide and run at people. He seemed to be taking everything out of my game that made it what it was. I was grateful for the times when I got to train with the first team, whose sessions were taken by Steve Coppell's assistant Ian Evans. He was a good guy and I enjoyed training with him. I think he rated me too and when I came to leave he tried to persuade me to stay. Perhaps he thought that with Wright and Bright likely to be sold in the next couple of years, I might have made a first-teamer and the club might have seen a return on their investment. I also heard later that when he was McCarthy's assistant at Millwall, he recommended that the club sign me.

With the confidence of youth I was not overawed by being at such a big club as Palace and reckoned I could be a good league player given the chance. I was impatient but realistic; I knew it was going to be a long time before I got past Wright and Bright. Palace had started the season well and were soon in the top three, even going top in September by beating Leicester 2−1 − how ironic that result looks today given what would happen at Wembley some years later. I could not see my way out of the reserves – for whom I came up against my old Weymouth colleague Peter Guthrie in a game against Tottenham. We also had in our team a teenager called Gareth Southgate. He used to play at right-back and even then was a level-headed lad who I thought would make it. I don't recall him taking any penalties, though.

Wright and Bright were excellent strikers, but both seemed big-time to me, and I thought their overwhelming influence was unhealthy. They could say and do anything they wanted, because they were the best players and they knew it. It seemed that all play had to go through them, was geared to them, and I had never been at a place where that happened. I suppose my view of the situation is coloured by my experiences there. Since then I have met Mark again and found him a really good bloke and everyone in the game seems to like Ian.

Some lads feel funny at a London club if they are not from the Smoke. I know Stan Collymore was later uncomfortable at Palace with all the cockney banter that left him feeling an outsider. But I think I gave as good as I got and never felt out of place. I got on reasonably well with most of the players. The one I remember most is Dennis Bailey, who was a born-again Christian and a great guy. In conversation one day, one of the lads asked him what car he drove. He said he had a BMW, but we had never seen it; he used to turn up in an old Volkswagen Jetta with every panel on it either scratched or dented. 'Where's the BMW, then, Den?' someone asked one day. 'It's in my faith,' he replied. 'Well,' said one of the lads, 'give us a fiver and I'll clean it for you every week.' I also befriended Chris Powell, a left-back who would later come on loan to Aldershot and went on to Southend United and Derby County. One day I said I was going to Brixton to buy a new pair of boots. He offered to come with me, saying that you needed to know your way around there. I was glad he did.

After about six weeks I had been fined £800 thanks to the Guildford bypass. This is bloody ridiculous, I thought. I can't go on like this. I decided to buy a flat in Wimbledon but no sooner had it all gone through than I was being transferred. The flat cost £76,000 and I never got to live in it for one day. I still own it.

Unbeknown to me, the Aldershot manager Len Walker had picked me out when I played for Palace reserves against theirs and had made some enquiries. He found out I had not settled and Palace, no doubt taking into account my record of punctuality, were only too willing to get £14,000 for me. I was willing, too, to go somewhere I was wanted, where first-team football looked probable, and so in October 1988 I readily signed for the Third Division team. Len seemed all right, another one who was a bit dour, but then we rarely saw him except on match days, as he left the day-to-day coaching mainly to his skeleton staff. He wasn't the best-looking bloke in the world and to make yourself look better, you always tried to be next to him in the team photo.

I received a £12,000 signing-on fee, payable over two years, and wages of £350 a week. Now I had never been great with money; either obtaining it, keeping it, or negotiating on my own behalf, and I think I was a bit daft. Being a free agent I could have originally asked for more when I joined Palace considering that the club saved a bundle on a transfer fee. Also, despite moving on, I was entitled to my second £5000 from Palace, not having asked for a transfer. They didn't deliberately avoid paying it, however. It was my job to ask for it and I failed to do so. A lot is talked about agents in the game, and how greedy they are,

(Above) **The two of us. Steve and sister Ruth in the early seventies.**
(Below) **Me and the boys. Captain of Lock's Heath Sovereigns (with the ball), in the garage at home.**

Apprentice boys. Under the eye of Portsmouth manager Bobby Campbell *(far right)* **Steve** *(seated second from left)* **signs on at Fratton Park. Behind him, Anne and Alan Claridge stand proudly** *(Portsmouth & Sunderland Newspapers).*

S MAIL
L EDITION

RUARY 19, 1983   No. 32,923 (Est. 1877) 12p

● Pompey apprentices chip away the ice on the perimeter track. From the left Steve Claridge, Darren Brown, Jason Pearce, Mario Walsh and Mark Davies. — Picture 1877-0

SCORES CHECK

Dunham fight to their win

Read all about it. The glamorous life of an apprentice footballer.

Local zero. *(Above)* **With Fareham Town during an unhappy spell** *(Portsmouth & Sunderland Newspapers)*. **Dorset days.** *(Below)* **In action for Weymouth against Runcorn and** *(over)* **Altrincham** *(Geoff Moore)*.

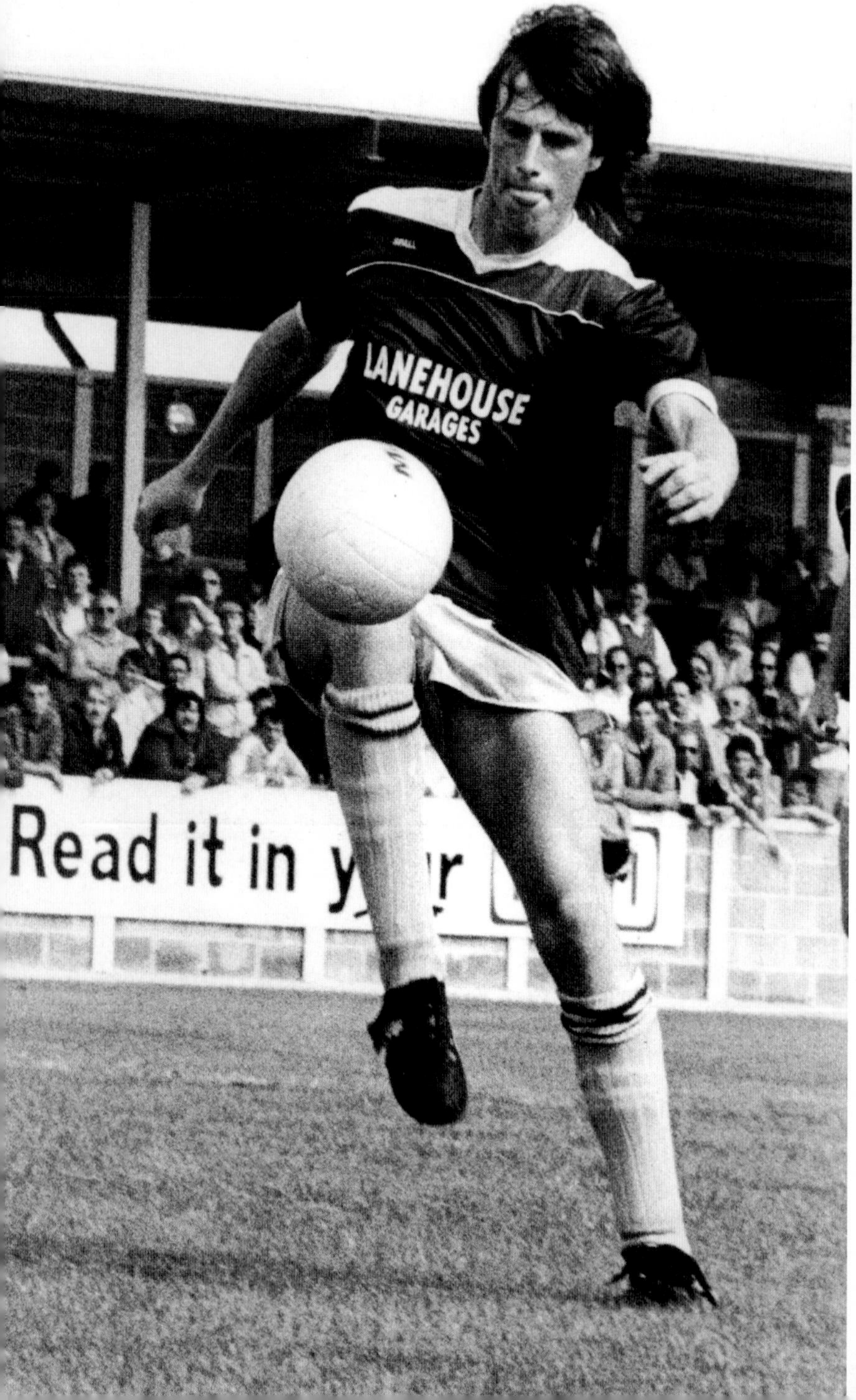
LANEHOUSE
GARAGES
Read it in y

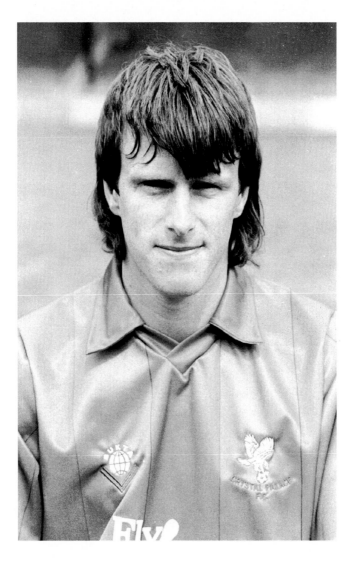

A rare shot in a Crystal Palace shirt *(Neil Everitt)*.

Shots of the shots. Signing for Aldershot, managed by Len Walker *(above)*; **sponsored by Lada at the Recreation Ground** *(below)*, **note the neat footwork to evade Exeter's Danny Baily.** *(Over)* **Don't look back in anger – in fact, fondness remains Steve's view of Aldershot.** *(Photos: Jeff Trolley)*.

Celebration days…having scored for (clockwise from top left) **Aldershot** (Jeff Trolley), **Cambridge** (Rob Gallagher), **Leicester** (Empics) **and Birmingham** (News Team)**.**

but sometimes young lads really do need guidance. Clubs will take advantage if they can. In those days few agents were interested in the likes of me, although today the Professional Footballers' Association are very good at helping out when it comes to talking terms.

So began what was one of the most incredible experiences of my life. The place and people were crazy and my period at Aldershot was to be one of the happiest of my career, despite all the things wrong with the place. I thought to myself quite early on that I'd finally found the right club for me. It was all just incredibly funny, a laugh 365 days a year.

It was also the sort of experience that stands you in good stead for the rest of your career, makes you appreciate it properly when life does improve. It made me realize that not everything comes easily and, as someone once said, that which doesn't destroy you makes you stronger. We were all earning a pittance, too – that was when Aldershot paid us – and for married men with kids and mortgages it was probably barely enough money to survive on. People get carried away with what footballers are earning these days but two-thirds of them are just surviving, not even earning what would be a decent wage in industry. You have to remember that every game could be a player's last and he would be left with no source of income. It's why I don't begrudge any player the best deal he can get, even if it is daft dosh. You have to make the best of the situation.

We certainly did at Aldershot. Things were not promising when I joined, I knew that, but I stuck with my philosophy of going where you are wanted. The previous season they had escaped a relegation play-off by a point, having the year before reached the nose-bleed heights, for them, of the Third Division in an amazing way. They'd finished 6th in the Fourth, nine points behind Wolverhampton Wanderers, who were 4th in the table, but then astonished everyone by winning the play-offs. They went and beat Bolton Wanderers 3–2 on aggregate then Wolves 3–0, 2–0 at home and 1–0 away.

Now they were firmly rooted in the bottom three of the Third and everyone at the club seemed to know from an early stage that this time relegation was a probability rather than just a possibility. My first game was quite an introduction. We had lost 5–1 at Fulham and were trooping off appropriately dejected. The walk back to the dressing room in the corner of Craven Cottage took us between a paddock of home fans and the away end and the first thing I saw was our midfield player Glen Burvill sticking two fingers up and shouting eff off to a group of about forty fans. Fulham's, I presumed. Getting nearer, I noticed they were ours. Because Glen had once been at Reading, Aldershot's hated localish rivals, he used to get more stick than the rest of us. Mind you, none of us was exempt.

Our goal that day was scored by Giorgio Mazzon, who was our midfield dynamo. A dynamo with a disabled sticker in his car, I found out early on after training one day. Giorgio had been a promising young player at Tottenham, and was still handy, but had been in a car crash and damaged his back. Now he had, despite being a professional footballer, a registered handicapped notice in his car, which allowed him to park almost anywhere. Only at Aldershot would a handicap not have been a handicap.

The place was full of characters. There was Steve Berry, a good pro who had been at Portsmouth when I was an apprentice. We used to travel up together, me having moved back there with Mum and Dad because the travelling was easier than it would have been from the flat in Wimbledon, him bringing his dog with him. We'd stop for the greasiest bacon roll you could imagine on the way up. Later in my career, after I left Cambridge United for the first time, Steve rang me to say he was playing in Hong Kong and that there were good opportunities there if I fancied it. Then, a few years later, I woke up too early one morning, having wrongly set my clock radio alarm, to hear a match report of a Stevenage Borough game with his name being mentioned as a midfield inspiration. I'm not sure it was the right description but it turned out to be him all right.

My partner up front in my first season was Dale Banton, who couldn't train because his knees were so knackered. He just used to come in and put ice on his knees. He had come from York on astronomical money after having had a good spell with the club previously. I don't know how he passed the medical. Come to think of it, I'm not sure they had such things as medicals at Aldershot.

We had, too, a young player called Darren Anderson, who but for a consonant in his surname might have gone on to play for England. He was a completely different build from Tottenham's Darren, though. With his blond crewcut, he looked like that old American footballer Keith 'The Boz' Bosworth.

There was also David Barnes, who, according to a rumour that he had no desire to deny, once joined with Billy Whitehurst in a bout of mile-high jinks while on a club tour with Sheffield United, forcing a pilot to land an aeroplane ahead of schedule and at an unscheduled airport. If Gazza is as daft as a brush, David was a cupboard full of them. I remember him in a night-club wearing a thick, heavy overcoat, refusing to take it off or hand it to the cloakroom staff because it had cost £500. He reckoned he looked cool in it 'because no one else had one like it'. Once he brought into training an ordinary chair with wheels on it, took it out on the pitch and pretended to be Giorgio Mazzon, with everybody falling about laughing. Luckily Giorgio took it in good spirit, but then that was the mood of the club at the time.

I didn't escape either but I enjoyed all the banter and the ribbing. One day my dad came to a game and I took him into the players' lounge after the match. Once he opened his mouth to reveal his farmer's accent I got the first of several nicknames I have been awarded down the years: Worzel.

Training was somewhat chaotic and something of a mess – literally. Sometimes we would be allocated a public park which was covered in dog excrement. It did improve your footwork, mind. Not surprisingly, the goalkeepers would refuse to train on it, preferring to steal one of the young lads for a game of football golf, which consisted of digging a hole, going back about 200 yards and seeing how few kicks it took to sink the ball. We had an Army ground for a while, but not often enough. It should have been simple enough in the town that was the home of the British Army.

Mostly we had to train on the car park at the Recreation Ground, which was usually half covered with rain water and had hardly enough space for a five-a-side. Again the goalkeepers would get out of it, but nobody tried too hard, not wanting to risk injury. Generally, training was a pretty lacklustre affair, often finishing early with a five-a-side game.

I rarely took a shower afterwards, not only because I was keen to sell some fruit and veg from the back of my car to the lads, which netted me about £50 a week, or wanted to get down to the bookie's, but also because the drains were all blocked and the tiles were off the wall in the dressing room. Besides, one player who was notorious for such things might have been on the prowl and done something unmentionable to or in the bath.

I was a bit in and out, up and down that first season, but I had my moments. In my fourth game I got my first goal, to give us a 1–1 draw at home to Chester City, and the following week, in early November, came what was to prove the highlight of our season. Sheffield United, gradually hauling themselves back up through the divisions, now under Dave Bassett, came to the Rec one off the top while we were one off the bottom. The crowd of just under 3000 seemed three parts theirs, but we scrambled a 1–0 win with my goal. As the ball came over it just seemed to hit my head on the edge of their penalty box and loop over their keeper. I can still picture it falling under the bar now as I was tumbling. Apart from that, I had a 'mare' of a game, as I recall.

We also had an FA Cup epic in the second round with Bristol City around that time, the days before penalty shoot-outs after the first replay, which I think undermine the whole spirit of the competition. We had to play four games in twelve days just before Christmas to resolve it. Three times in three matches we took the lead before conceding an equalizer, at least once in the last minute, and we would end up drawing. Finally in the fourth match Carl Shutt got a winning goal for them. Aldershot would not be going to Wembley that year.

That Christmas the club told us we would all be getting a bonus. A night out or a few quid extra in the pay packet, we thought. Then on Christmas Eve it arrived: a turkey and a £2.99 bottle of Mateus Rosé. I took them home to my mum, all the while on the journey back wondering what the funny smell in the car was. When she unwrapped the turkey, she found it was rotten to the giblets. Luckily we were not relying on it for our dinner.

If that didn't sum up the state of the place, what did was the car sponsorship deal we were offered by a local garage. We didn't expect Mercedes, Rovers or Escorts, perhaps, but the best the club could do was four Lada Rivas, horrible box-shaped efforts which cost about the same as a ten-year-old Mini. When they turned up, only laughter hid the disappointment. I think one or two of the older players, the impoverished ones, agreed to have one but no one under 45 would be seen dead in one. Later that season Aldershot took a Russian, whose name escapes me, on loan and I'm convinced it was just to get rid of another of those Ladas.

Never once did we get out of the bottom three in the New Year, although we threatened to improve with the signing of David Puckett from Southampton to play alongside me up front. We reached the dizzy heights of 22nd in early February with our best win of the season, 5–1 at home to Northampton with me scoring and David, a really quiet, family man who was a bit out of place at Aldershot (he used to travel home with me in the car sometimes shaking his head for the whole journey at what went on at the club), getting a hat-trick. I ended up with 10 goals from 41 appearances for the season, second top scorer behind David, which earned me the supporters' vote as Player of the Year. It really touched me. The award was presented to me by the club's president, the comedian and former cleaner in *Are You Being Served?*, Arthur English.

We were duly relegated as bottom club and in some ways it was a bit of a relief as we had been struggling out of our depth, getting tired of losing. Perhaps the Fourth Division would be more our level. Hardly. After pre-season training at a local point-to-point course, which featured players trying to leap the fences for a laugh, we lost three of our first five games, drew the other two, scoring only one goal (a penalty), and by mid-September were bottom of the Football League. Then, for some unaccountable reason, we clicked, winning four games in a row, drawing the next two and moving up to 10th, the highlight for me being my hat-trick in a 4–2 win over Scunthorpe United, the first of my career in the league. We also amazed ourselves by overturning a 2–0 deficit in the first leg of a Littlewoods Cup first-round tie against Peterborough United, winning the second 6–2 to earn a second-round shot at Sheffield Wednesday, then under Ron Atkinson and expected to be one of the best sides in the country that season.

In fact they were terrible in the first leg and we got a 0–0 draw at Hillsborough. They had a lot of possession but few ideas and created very little. They were worse than anything we had faced so far in the Fourth Division and, in fact, I went out and backed them to go down after that, though in the end they just avoided it. We really thought we had a great chance in the second leg on a big night with a crowd of 4011, our biggest of the season, in the Recreation Ground. We lost 8–0. And it could have been 18. Appropriately, at the home of the Shots, this was shooting practice. It was embarrassingly one-sided. I think we were four down after quarter of an hour and six down at half-time. In the end Steve Whitton scored four and Dalian Atkinson three. I remember counting our side because I didn't think we had enough men on the field – when I wasn't looking at the bench hoping they were holding my number up to be substituted, that was.

Around that time the rumours started flying that the club was in financial trouble. Rumours soon became reality when Aldershot announced that they could not pay us, a situation which went on, first time around, for nine weeks. We had a players' meeting and Len Walker came in to tell us that we didn't have to train if we didn't want to. We didn't want to. You would be spending your own petrol money – £50 a week – with no chance of getting it back from the club. Not being paid was one thing, paying them quite another. We still turned up for matches, of course, because it was the only chance you had of getting paid anything. Besides which, when you are a player you simply want to play.

The fans rallied round, starting an SOS appeal – Save Our Shots – as well as putting on several functions, and somehow we kept going. We used to be concentrating more on the collecting buckets going round the ground than the ball while we were having a pre-match kickabout because it depended on how full they were as to whether you would get any money or not that day. Still, I don't think anybody realized how serious it was. When Aldershot eventually folded eighteen months later, in the 1991–92 season, the debts were around £1.5 million. How a club that gets fewer than 2000 through the gate is allowed to run up that amount of credit is beyond me. It wouldn't have been allowed with a bookmaker, though it showed me what damage I could have done if I had been indulged.

Of course the lack of training affected us as we gradually slipped down the table. The odd good result interrupted it, like a 6–1 home win over Hartlepool in which I got a hat-trick, and 2–0 over Halifax, where I scored both. A few days after the Hartlepool match I went into the club shop to buy the video, and when I got home I put it in the machine. I didn't expect John Motson, but the commentator was terrible. After about 15 minutes I recognized him – the coach driver who took us to away matches. He was getting all the names wrong. Mostly all I could hear though was some bloke up in the stand, obviously next to the commentator, bellowing: 'You useless prat, Burvill.' It was all filmed on one camera, which zoomed about all over the place, mostly missing the ball, and wasn't edited at all. Still, at least my hat-trick was recorded for posterity, and it would later prove useful in securing me a transfer.

Our morale was getting lower as the club bounced numerous cheques on us. Most clubs pay you through the banking system but Aldershot insisted on writing you a cheque. It was not that the club could afford expenses, either. We never stayed overnight on an away trip; if it was Carlisle, we would just have to be at the ground at 6 a.m., and get back in the early hours of the next morning. By the end there was no coach, and the players were asked to drive their own cars. The club secretary Steve Birley, who lived down in Exeter – I think he was the only man willing to take on the job – also started sleeping in his office on a camp bed to save on bed and breakfast bills.

Some players were in danger of having their houses repossessed when, thankfully, the PFA stepped in and paid us. We also had another period later in the season when they baled us out after another six weeks without money. The problem was that the club never really had much potential, charming and pleasant as it was with its tidy little ground – though even this was falling into disrepair, as one set of away fans found out when a section of a stand roof fell in at one game, luckily without injury to anyone. (There weren't that many in the ground, anyway.) And floodlight bulbs that had failed would not be replaced. Still, the floral gardens at the High Street end of the ground, which meant the club could not have any terracing there, were quite pretty.

If Aldershot had been a sleeping giant, I am sure somebody would have taken it over and invested in the club. After my time there several people had a look, including a consortium fronted by the former Ipswich, Arsenal and England midfielder Brian Talbot. And one publicity-hungry young property tycoon called Spencer Trethewy also declared an interest – even going on the Terry Wogan TV show to talk about it – but he proved not to be genuine. The problem was you were never going to get more than a few thousand through the gate so there just wasn't the potential there. With a population of fewer than 40,000, Aldershot was at that time the smallest town in England with a league club. It may have been famous for the Army, but there was little interest in this outfit of small Shots.

Despite all the problems, nobody ever considered refusing to play for the club. At that level you are not in it for the money anyway. Most of us were just in love with the idea of being a professional footballer. Strange as it may sound to the cynical, footballers do like playing most of the time. You live for your shot at glory, that one victory, that one Cup result that lifts your name out of the small print and on to the back pages. And Aldershot was such a fun club, with a great collection of players bonded by adversity, that these remained enjoyable times with or without the money.

But hard times as well, and whenever they got an offer for a player, he had to go. Soon it was my turn. In November, we had gone up to Cambridge United and got a 2–2 draw after being 2–0 down. I thought I had played quite well against Liam Daish, who really was a tough stopper of a defender, difficult to get the better of. They obviously remembered me and had also sent someone to watch me at the Hartlepool game, who, like me, studied the video afterwards. They were building a decent side by the look of things. John Beck – who knew me from Bournemouth, of course – had just taken over as manager and offered £75,000 for me, which was a huge lifeline for Aldershot.

Len Walker informed me the offer had been accepted and said I should travel up to discuss terms with them. I was still very naïve then. If I had stayed at Aldershot until they folded, I would have been entitled to a free and could have negotiated a big signing-on fee with a club. But I would actually be doing Aldershot a favour by leaving them. And this was a chance to get paid again.

I was determined to take it when I drove up to Cambridge in early February of 1990. New decade, new start. But it was to be another of those inauspicious starts I seem to make with clubs. And after the relatively happy times I had been used to at Aldershot, the move to Cambridge was to be the start of the worst two years of my life and my career.

5 Cambridge Blues

The Den in those days was one of the most intimidating
venues in the country for visiting teams. The atmosphere
around Millwall's crumbling ground in the rabbit-warren
backstreets of the tough New Cross suburb in south-east
London was best captured by one of their own. Eamon Dunphy
was a talented ball player from the Republic of Ireland in the
seventies, who moved on from the unpromising position of
the team's midfield – though there were signs in his game of
the lyricism of his Celtic background – to become one of
football's most stylish, incisive and outspoken writers:

'We always expect to win at home. Teams hate coming to
The Den, which is partly why we beat the Football League
record for being undefeated at home,' he wrote in his brilliant
diary of a season, *Only a Game?*, in 1973–74. 'When I first
went there I used to hate it. I remember going there with York

City for my first visit. It took us half an hour to find the place. Eventually we went up this dingy backstreet. I remember thinking, Where is this? Then you go and have a look at the pitch, which is bumpy, terrible. And you think, Oh, Jesus. What are we doing here? The dressing rooms are terrible, small poky places. The away dressing room is a dungeon, no light, no window. The bathrooms are horrible. Then you get out there to face them – the Lions. And they come storming at you. Most sides jack it in.'

The Den was also one of only a handful of English grounds with the dressing rooms under one end of the stadium rather than in a main stand. It was, too, at the end that housed home fans, whose stamping and shouting could be heard on the ceiling of the 'dungeon'. I wondered what it would be like for players about to enter the Lions' Den, how unnerving it would be as the sinews in the stomach rearranged themselves into knots.

In January of 1990, in the fourth round of the FA Cup, Fourth Division Cambridge United were drawn at Millwall, who were then riding high in the First Division, having topped it earlier in the season. Having recently read again Dunphy's seminal book, I telephoned Cambridge's new manager John Beck to ask if it might be possible to spend the afternoon in the dressing room to find out first-hand just how fearful an experience it was. Millwall had plans to move to a new state-of-the-art, all-seater stadium a mile or so away and I wanted a flavour of the English game's traditional culture before progress caught up. To my delight, Beck readily agreed. 'I think you'll find some of the things we do very interesting,' he promised.

I did. When I arrived on that crisp winter's day and was introduced to the team, they struck me as a young, impressionable and happy bunch who clearly believed that they were in on the start of something. Though mid-table in the Fourth Division, the appointment of the youthful Beck, then 36, with the previous manager Chris Turner having moved upstairs to become general manager, had clearly lifted the team. 'Come and watch this,' he invited, as I sat in the corner of the dressing room, which was indeed an inhospitable place, though modernized since Dunphy's day. It was airless and cramped, being half the size of the home one, and the rumble of the crowd insinuated its way down through the ceiling panels, growing noisier and more threatening as kick-off approached. The Cambridge players joked about having a ballot to determine who would have to take the corner kicks.

Round the corner from the changing area, past an axe in a glass case for use in case of fire – or perhaps to ward off invading hordes – a vicious shower jet was spewing freezing cold water. Each player, except the substitute, was required to spend ten seconds – counted out by team-mates – under it. As they emerged, the substitute, Gary Clayton, was enjoying the task of throwing a bucket of yet more icy water over them. It was designed to wake them up. It could hardly have any other effect.

Beck's team talk was high on motivation. 'I've been telling everybody you lot have got arsehole,' he told them. 'Now go out and prove it.' Then all locked arms in a show of solidarity and bellowed: 'Let's go.' Over ninety minutes they did prove they had the commitment and resilience that Beck meant, I think, by 'arsehole'. A goal by Tony Cascarino, playing alongside Teddy Sheringham, gave Millwall the lead but the leggy, lanky striker John Taylor scored an equalizer in the second half.

It was a richly deserved draw and an enriching experience to be on the inside of that day. The players were a promising crew: the barrel-chested John Vaughan in goal, the quick overlapping full-backs Andy Fensome and Alan Kimble and the solid, capable Phil Chapple and Liam Daish centrally forming a formidable defence. Michael Cheetham and Lee Philpott were swift wingers who crossed a mean ball for Dion Dublin, wearing the newly trendy cycling shorts for the first time that day to the ribbing of the team's wit Gary Clayton – 'It's only because we might be on *Match of the Day*. You look like that runner, Tony Christie,' said Clayton. I fell for it, to hoots of derision from him and the team. 'Don't you mean Linford?' I asked.

Alongside Dublin, John Taylor completed a potent partnership up front. Linking it all was a central midfield of the workmanlike Colin Bailie and the thoughtful Chris Leadbitter, with his sweet left foot.

Into such an atmosphere stepped Steve Claridge a fortnight later on February 8, signed from Aldershot for £75,000. Cambridge had just beaten Millwall 1–0 in the replay to wrest some national attention from the bigger clubs. All of a sudden interest in Beck's methods, notably the cold-shower shuffle, took off and we had another of those charming Cup giant-killer stories on our hands. Those of us who had known Steve and noted the transfer felt that he had landed on his feet.

Actually, his feet were never to touch the ground. I'd personally had an indication that this was not merely a romantic adventure when I telephoned Beck after the Millwall victory to offer my congratulations and seek his comments for a follow-up story for my then employers, the *Guardian*. 'Any money in it?' he wondered. Soon Steve would be seeing at first hand that Beck had anything but romance in his soul. If football was a hard game, Beck was determined to make it harder.

Steve didn't help his own cause early on, turning up late for his own signing, as his habit of insouciant punctuality was repeated. 'I remember about two days after he signed,' recalls his team-mate Phil Chapple, 'we were in the middle of a team meeting and Steve's head poked round the door. Becky wasn't best pleased.' Neither was the manager impressed when Steve refused to enter into the spirit of the cold showers. 'It started off as a bit of a laugh,' says Chapple. 'After the first time it was suggested our form improved, so we carried it on. The worst places were Carlisle, where it was one of these powerful overhead showers rather than one that came out at your chest, and Scarborough, who didn't have showers so we had to have a cold bath. Anyway, Steve said he wasn't doing it because of some heart trouble. We all shouted, "Make him do it," but he got out of it. He was good at that.'

Chapple also had first-hand experience of the Claridge animal in its habitat. 'Chris Turner put him in digs with me at first,' he recalls, a mixture of amusement and annoyance in his voice. 'I would go to bed at about ten-thirty and be asleep when Stevie came into the room, a bit noisily, at about one o'clock. He would then turn the light on and read, waking me up. After I got back off, I would wake up again at about three to find the light still on, the book on the floor and Steve asleep. It only lasted two or three days. Steve moved out. He was fussy about his food and the way it was cooked. I mean, it was normal food, but he just didn't like it.'

Chapple recalls, too, an end-of-season club holiday in Spain. 'Everyone turned up at the airport with a suitcase,' he says. 'Steve had a tiny holdall. He lived in the same clothes for a few days until Gary Clayton got sick of his stuff lying all round the place and threw it all, holdall included, on to the roof of the place we were staying at. Steve just left it up there until fifteen minutes before the bus was due to leave to go back to the airport on the Friday then went up to fetch it. We were all sat on the bus in fits of laughter watching him scrambling around trying to retrieve it.'

Chapple and Co. were quickly aware, too, of Steve's

gambling. 'We used to come back from training and the first thing he did was run across the road to get a bet on. He never hid it, really. Like most people who have a problem with it, they are quick to let you know when they have won. It's not quite the same when they lose.'

John Taylor was once asked by Claridge to pick up some winnings from a little bookmaker's down a backstreet near the Abbey Stadium. 'While I was in there, there was a dog race on and I took £20 out of Steve's winnings, which were a few hundred quid, and put it on a dog I fancied,' Taylor recalls. 'It romped home at 10–1 and I picked up £200, replacing the £20 in his winnings before I handed his to him and keeping mine. I can't remember if I ever told him that. I was a bit of a gambler but not in his league. He was a bit over the top for me.'

After his unpromising start, Claridge repeatedly clashed with Beck as the style of playing evolved into an ugly, long-ball game devoid of any subtlety. It may have seemed at odds with the image of Cambridge as one of the world's great seats of learning and intellect, its college architecture evoking English civilization at its haughtiest, but then town and gown were ever disparate. United's home at the Abbey Stadium was on the edge of town, divorced from the university's supposedly more gentlemanly pursuits of cricket and rugby union at Fenner's and Grange Road, and well away from the rowing on the River Cam. Out on the Newmarket Road, the bleakness of the spot was interrupted by allotments and industrial estates. *University Challenge*? With Claridge it was more like compatibility challenge.

'Steve was an individual and Becky was all for team things,' says Chapple. 'It wasn't for the purist, lots of one-touch, hitting the ball for the corners, set pieces, first-time crosses. If you disobeyed him you were off, whether it was after twenty minutes, as I remember happened to Steve in a friendly against Cambridge City, or eighty-five. We were mostly all young lads and we went along with it because it brought us success. Also, you didn't want to lose your place because there were always people in the background waiting to take it.'

Claridge suffered the most but others suffered too, and curiously he benefited on one occasion. 'Becky really motivated us but Steve was having none of it,' says Taylor. 'I also remember saying something about the management once. I was dropped, with Steve taking my place, and I couldn't get back in for four months because Steve took his chance and played really well. In fact we had sponsors who awarded hi-fis for man of the match and Steve won three. After that, we agreed they should be sold and the money divided up amongst the squad, though one player welched on that.' Claridge and Taylor, who was eventually sold to Bristol Rovers, remain friends despite having competed for the same place.

For a couple of years Beck's methods did indeed bring success. Cambridge soon got out of the old Fourth Division, went through the Third and took the Second by storm. Football had another Wimbledon on its hands. No sooner promoted, the U's led the table early on and were still in the promotion frame at Christmas. Cambridge's fanzine, neatly titled *The Abbey Rabbit*, was delirious. Surely this mottled crew were not going to be inaugural members of the new Premier League, designed for the big clubs and their bank balances, in the 1992–93 season? Ultimately not. The game's governors heaved a sigh of relief as Cambridge faltered in the second half of the season, eventually thrashed by Leicester City in the play-offs.

'The problem was that after we had played everyone once they knew us inside out for the returns,' says Chapple. 'At that level, you had a better class of coach and player, who were able to adjust their game to counter us. It was a bit easier away from home where teams have to come at you, but at home we couldn't break them down. We would have most of the play but didn't threaten much. It wasn't too bad for the lads in defence but we felt sorry for the forwards, and we had three good ones in Dion, John Taylor and Steve. We asked Becky to give them more licence in the final third to do things on the ball, but he wouldn't have it.'

Claridge was one of the most outspoken within the team about the tactics and was frequently used only as a substitute, though his scoring record remained good whenever he did appear. 'Stevie did clash with the manager a lot and he was sub more than anything, but he always did a hell of a job for us,' says Chapple. 'His work rate was phenomenal and you knew as a defender that if you looked up you could hit him with the ball with confidence. The fans also loved him. He was incredibly fit.'

Was that not surprising given Claridge's diet, training regime and lifestyle?

'Yes, I suppose it was, because he didn't eat that well. He loved his fruit and chocolate. But it was just natural, I think.'

'He would only ever eat chicken, with no sauce and a few vegetables, when we stayed away in hotels,' says Taylor. 'Then he would always take the bowl of fruit from the table up to his room. He would also strip the hotel bed and put on his own sheets and pillowcases, which he brought with him.'

Steve's relationship with Beck came to a head as Cambridge's challenge for automatic promotion stumbled during a match with Ipswich Town, at half-time of which he and Beck were involved in a fight. 'Yes, I remember it,' says Chapple. 'I was in the dressing room and this was going on in the medical room where Steve had gone to escape from him.

115

I just heard a furore and then some of the lads went in to break it up.

'Becky did take us a long way, to be fair to him, and he always said it was his system, which just got more extreme. I don't know if he might have mellowed now, but we also had some pretty good players there, as was shown by a lot going on to bigger clubs.' Chapple himself went on to suffer a culture shock when he moved to Charlton Athletic and became a part of their passing game. How green was the grass at the Valley.

'You could say that the manager and Steve didn't see eye to eye. They were just different personalities. You couldn't help but like Steve. He was a lovely fellow. You couldn't rely on him to do something for you, but I have stayed in contact with him, which doesn't always happen when players change clubs, so I suppose that shows what I think of him. There's only one like him, that's for sure. He lives life on the edge. Very eccentric. But then, if you took that away from him, what have you got left?'

'Are you taking the piss? Have you come on your bike or something?' were the first words Chris Turner greeted me with when I arrived in his office at Cambridge United's Abbey Stadium. I was a bit late, I admit, but only by an hour or two. My dad's old X-reg Cortina had broken down on the M11. Not broken down, exactly. It ran out of petrol. There were no services on the motorway and I was only one junction away from Cambridge. All the journey I kept thinking about pulling off, but then I thought I would just about make it. You know how you do.

I just left the car on the hard shoulder, knowing that no one would want to nick it, and cadged a lift from a bloke who was good enough to drop me outside the ground on the Newmarket Road. Then I shot in, out of breath, and made my apologies to this great big broken nose standing in front of me. Chris Turner had just moved upstairs to become general manager, with John Beck having taken over the running of team affairs, and the two of them seemed to have a Mr 'Orrible–Mr Nice Guy, bad-cop–

good-cop routine going. Beck seemed the decent sort he had been at Bournemouth. Yet again, first appearances were deceptive. I, and football, didn't know what we were in for.

One of the Aldershot lads, Colin Smith, to whom I will always be grateful for the advice, had told me to ask for a signing-on fee of £25,000. I was reluctant to do so, thinking I would never get it, but managed to spit out the figure as we were talking. To my amazement they agreed, phasing it over a two-year period. It seemed an incredible amount of money to me at the time. I was also going to get £450 a week, a rise of 33 per cent. Even though Cambridge, like Aldershot, were in the Fourth Division, they seemed to be a club on the up and I thought I was really starting to get somewhere in the game. I signed the same day, a Thursday, as Cambridge wanted it done and dusted so that I would be available for Saturday's home match against Exeter City.

I improved out of all recognition by then, turned up on match day at the appointed time of 1.30 p.m. and was named in the team. Then, as I was getting changed, Beck noticed that my boots were dirty. 'You'd better change those. No player of mine goes out with dirty boots,' he said, and threw them away. At that time I used to carry a bag full of old boots, about twenty I think, all in various states of filth and disrepair, and I searched through them for a pair that might be acceptable. None was, and Beck promptly dropped me from the team.

My explanations were not accepted, but I still think they were valid. I know I have a reputation for being a scruffy sort of player, but there is method in the madness. I carried the bag because I would change my boots regularly, depending on my form, although I frequently wear the Puma brand despite not having a boot deal with them. Some of it was superstition, the rest of it because I needed to feel comfortable in my boots, which are, after all, the tools of my trade. Confidence is everything for a footballer and I had to be confident in what I was wearing. Gary Lineker sent home for a pair of his old boots when he was in Japan.

I never let apprentices near them to clean them as I hate that shiny feel to them. They also use a wire brush which gradually wears the leather away. Also, at Aldershot, an apprentice once forgot to pack them for an away match. So at that point I never let them out of my sight. I always felt my touch was better with worn-in boots and sometimes I would even wear an odd pair. I have been known to keep the team waiting as they were ready to run out while I searched for just the right blend. Nowadays I am a bit more relaxed about it, having come round to the view that it may just be me that's crap sometimes, and will wear newer, cleaner boots, but in those days I preferred the lived-in feel. Psychologically it was a big part of my game.

I also dislike wearing clean socks. With these modern nylon ones, you just slip about inside your boot. To this day I still prefer to put them on with the sole scuffed, and after a match I will gather them up, along with my boots, and pack them in my kit bag to bring along for the next match in a used condition rather than allow them to be washed. They also shrink when put through a machine. It may sound strange, and perhaps a bit unhygienic, but it suits me. I also can't be doing with shinpads, which make me feel more cumbersome than I may look. Some referees have booked me down the years, and these days, due to FIFA rules, some insist I wear them, so I have to.

None of this cut any ice with Beck that day against Exeter, although he later came round – in despair, I think – and I didn't even make it to his celebrated cold-shower routine that was a feature of pre-match preparation. My new colleagues had told me all about it, though. 'Right, cold shower time,' Beck would announce an hour before kick-off. This was an idea conceived by Beck's assistant Gary Johnson at an away match earlier in the season when he thought the players looked a bit sluggish as they got off the coach. It was quickly latched on to by a manager who was fond of any old Iron-Man routine that toughened up the players. It consisted of standing under a freezing jet while the substitute counted out ten seconds then threw a bucket of icy water over you as you emerged blue and shaking. Some of the lads seemed actually to enjoy it: when I was later at Birmingham City with Liam Daish, he was still doing it as part of his pre-match build-up, but it was never my thing.

Despite getting off on the wrong foot with Beck for the Exeter match, which we won 3–2 after being 2–0 down at half-time, he gave me the no. 9 shirt for the next match against Southend and I managed to escape the full cold-shower routine, dipping myself in and out quickly when he wasn't looking. Before the next game at Gillingham he caught me doing the same thing, however, and insisted I go back for the full Monty. 'I'm not doing it,' I protested. He asked me why not. Telling him it was early March and that I was cold enough already, thank you, did not seem to work.

Then, thinking on my feet with a skill I had acquired at school, I mentioned the heart condition that had been detected as a 12-year-old but which, thanks to the medication, had never given me any trouble. 'I need that in writing from your doctor,' Beck replied. 'In the meantime you're dropped.' Because it was an away game and the whole squad was not present, I was made a sub. At home I would have been out altogether.

'I'll sort this, no trouble. You watch,' I said quietly to one of the lads. And the following Monday I went to a doctor and asked him to write a 'sick note' along the lines of 'This man must be spared any sudden shocks due to a long-standing heart complaint.' I think the quack was quite used to, and weary of, all the excuses he was forced to issue and didn't ask too many questions. And he did have my medical history. Beck also had to accept it, albeit reluctantly, when I took it proudly into the club. Mind you, from then on I was the one throwing the buckets at the other players. It seemed I had come out of that particular deal pretty well, though I must admit to wondering whether I could rub out my signature on the other one after the start I had made with this sergeant major of a manager who promptly kept me as sub for the next three games.

To be fair this was probably due to him wanting to retain the same team that was enjoying a good FA Cup run, for which I was ineligible – having played for, and lost with, Aldershot against Cambridge United in the first round. Cambridge had beaten Millwall then Bristol City 5–1 in a second replay. By now it was being confirmed how physically hard it was going to be. Although Cup-tied, I travelled with the squad. The pre-match accommodation was an Army camp in the West Country some-where. Cambridge would go on and give a good account of themselves in the sixth round before losing 1–0 to Crystal Palace, who went on to beat Liverpool in the semi-final but lost the Final to Manchester United in a replay.

Perhaps all the national publicity and success Cambridge enjoyed that season deflected attention from what was really going on, or perhaps Beck felt it was vindication of his methods, but away from all this romantic giant-killing stuff, things that were happening at the club added to my doubts about the wisdom of having joined. Beck's rigid regime started to get more and more extreme, started to go crazy, in fact. Some of the things that were done and developed over the next two years were just not acceptable to my way of thinking. They went way beyond any debate about the worth of the long-ball game.

Beck had the grass grown six inches in the corners of the Abbey Stadium's pitch so that the ball would hold up there when we knocked it into the corners, as we frequently did, to stop defences trying to pass it out. Signs were put up in the corners urging 'Quality' but to me there seemed little class about banging the ball down there. He used to put the baths on in the away dressing room before a game to steam it out and would put pounds of sugar in the visitors' tea with the aim of slowing them up. The opposition were given dodgy balls for the kick-about. During the week, we used to train on the pitch to rough it up so that anybody who tried to play football against us would be at a disadvantage. We just used to boot it upfield, so it didn't matter to us. The ball boys were given towels to wipe the ball before throw-ins for the home team but told not to do it for the away team. They were ordered to return us the ball quickly but slowly for the opposition.

Beck also had all sorts of sayings and pseudo-philosophy he used to ply us with, along the lines of: 'It's not the size of the dog in the fight that counts but the size of the fight in the dog,' mostly stuff I have long since banished from my mind, I am relieved to say. He brought in statisticians with crazy charts to lecture us about the efficiency of long-ball football and how few passes you needed to score a goal. He would also award weekly bonuses of £15 to the player whom he deemed to have played the system best or who kicked the ball the furthest. I hardly need to say that I never won a penny. It was like being part of a cult, all designed to convince us that the Messiah's way was the only way.

And for a while I suppose it did. That first season I was there, we slumped in the league during the Cup run, but once eliminated we picked up again, which merely showed to him, I think, that his medicine was doing the trick.

He gave me a run in the side for the last twelve games of the season, and we won eight of those to finish 6th in the table. I finished with 4 goals. It took us into the play-offs; ironically we confirmed our place with a 2–0 win in the final game over Aldershot, who had slumped badly as all their financial problems continued and finished 22nd. That game was notable to me for the lumps that got kicked out of my legs. After the buffeting, I was talking to some of the Aldershot lads in the bar and they told me they had all put a tenner into a pool, which would go to the bloke who had kicked me the hardest. My old friend, or should it be ex-friend, Ian Phillips scooped it by a long way. I can still feel it to this day when I think about it.

We looked to have blown our chance in the first leg of the play-off semi-final when we only drew 1–1 at home to Maidstone United but bounced back to win the second 2–0 after extra time. My substitution by John Taylor was a sign of things to come, though. I was back to substitute for the Wembley Final against Chesterfield, to my great disappointment, but I did manage to get on and make my first appearance at the stadium; for the last twenty minutes.

I thought it was going to be a personal disaster, as I only touched the ball twice and one of my old split boots came off, taking about five minutes to get back on, but I do claim some credit for our 1–0 win. At corners, everyone was supposed to know where they should be standing. As we prepared to take one, I saw that Dion Dublin wasn't in the right position and shouted to him. He duly took up his appointed place and when the ball came over, he was on hand to score the winning goal.

The system was based entirely on set pieces. As a forward, you were not allowed to pass the ball backwards to lay it off for a team-mate, you were expected to lob it on regardless as to whether a colleague was on hand. If it went out for a throw-in, Beck still considered it a quality ball. We would just push up on the opposition to pressure them, they would throw it in and we would look to get a throw-in or a corner for ourselves out of it. Then we would lump the big lads up and hopefully score from a long throw or a corner.

To be fair, some of what he did was brilliant. The midfield and defence were so well organized, both operating in harmony as units, that they were excellent to play in front of. But you can afford to be predictable in those areas because you are not asking people to win you games from there. It was in the final third of the field that Beck's methods really grated with me. You just can't, or shouldn't, legislate for forwards. They are a different breed of player; you have got to play more off the cuff in the final third. In coaching a defender you can drill them to get it and play it down the line, but you can't say to an attacker, run down the line and cross it. It doesn't work like that. He has to have some freedom to use his skill to unlock a defence. It's what fans like, that inventiveness up front, and it is what gives the game its excitement. Fans love forwards trying to do things with the ball.

My reservations were confirmed at the start of the next season when we played a friendly against Cambridge City, the Southern League side in the town. I was a substitute again and eventually got on with ten minutes to go. I lasted four minutes before Beck brought me off. My sin was in getting the ball wide, cutting inside and passing it to Chris Leadbitter for a shot which went over the bar. I was supposed to knock it down the line into the corner for someone to chase, so that we could get a throw-in and launch a long one into their penalty area. He told me he would rather play with ten men than have me doing something I wasn't supposed to.

I started the season as a sub again, coming on for John Taylor, but we lost 1–0 to Birmingham City. The team's poor start, though, was enough to get me a run in the side and I started the next five matches. Still we were just not gelling and after a 2–0 win over Fulham, in which I scored, we lost the next against Southend and drew the following three. We also went out of the Littlewoods Cup to Walsall. It was back to the subs' bench – if I was lucky. Over the next six months I started only three matches, and was sub thirteen times on the occasions when I was named in the squad. My relationship with Beck deteriorated further. It was clear I was never going to be his cup of tea, as shown in October when I came on against Brentford, scored twice to open up a goalless match, and was back on the bench for the next game.

People may say I should have done what the manager wanted, that that was what I was being paid for. But I felt all the things that made me the player I was, the one he had bought, were being taken away from me. My game is based on first touch, holding the ball up and knocking it off. Not getting it and hoofing it into the corners. Training sessions often comprised a full eleven-a-side match of one-touch, the most boring ninety minutes you could imagine. It was possible only to head the ball out wide or flick it on; you were never allowed to play it back. If I failed to comply, Beck would make me wait and watch in the centre circle and send everybody on a lap of the pitch. I suppose he thought the team would moan at me loud and long enough so I would do what he wanted, but I'm afraid I didn't see it as much of a punishment. I confess to finding it funny that there was I sitting on the centre spot watching everybody else put in the effort. I do love hard work. In fact I could watch it all day. Besides which, I believed in a way of playing the game and this was not it. It was joyless and nasty.

I was singled out and I went about three months without getting a day off. A typical week would see me training Monday and Tuesday with the first team, sub for a match in midweek, then Thursday with the reserves, Friday back with the first team and sub again on the Saturday. On Sundays I would have to come in on my own, my only company some poor apprentice who was probably also being punished. We would simply have to run round the ground for an hour.

Perhaps Beck felt I needed the work. One Monday morning, after a rare Sunday going back to Portsmouth, I set off back to the club with just a bowl of cornflakes inside me. Feeling peckish, I decided to stop on the motorway for a full fry-up, followed by two bars of chocolate. After two more bars in Cambridge, I realized my greedy mistake. That morning we were sent on some hill running. I have no idea why, never having run up a hill during a football match (apart from at Wycombe's and Yeovil's old sloping grounds). But by the end of the run I was absolutely knackered and I flaked out. When I came round, I picked myself up to see the minibus that had brought us disappearing down the road. I then staggered out on to the road, soon to be confronted by two policemen, who told me someone had reported a drunk wandering about the streets. They seemed to accept my explanation when at this point a van arrived and the driver offered me a lift back to town. I was sat in the back surrounded by stinking paint.

Beck's schemes got dafter. I remember him bringing in a hypnotist once. For some reason he had us putting on tin hats – more useful for the opposition the way we played – and pretending to ride horses. Most of us didn't go under but four or five did. Our midfield player Danny O'Shea really got into it. He was sat on this chair, which he thought was a horse, and the bloke had him imagining he was riding in the Derby. He was totally gone and everyone was just roaring with laughter. He was really riding a finish, riding for his life. I have never seen a chair move so quickly. I just wish he had been on some of the horses I have backed down the years. I couldn't take it seriously and wasn't hypnotized, though I think I would have been the one Beck would have chosen to be, so I could be brainwashed into playing his way.

Because he didn't have a great depth of squad, Beck would name me as sub and stick me on if we were struggling. In fact, I warmed up at so many grounds when I was with Cambridge that I was thinking of bringing out my own exercise video. It worked for Joan Collins, after all, and she didn't have the range of stretching exercises that I did, despite the evidence of *The Bitch* and *The Stud*. In November, I did get a run of three games in the side due to injuries and scored two goals but I was soon back in the twilight zone. So too was the team; 13th in the Third Division table that month.

They put together a run of six wins in December and January, though, to move up to 3rd and also embarked on another good FA Cup run. Wins over Wolves, Middlesbrough and Sheffield Wednesday took them to another sixth-round tie, this time a real biggie, against Arsenal at Highbury. Although I wasn't expecting to be in the team as Dion Dublin and John Taylor were playing so well together up front, I was hoping to be a sub. Instead, I was dropped completely from the squad. I was told in the dressing room that Beck had decided on two defenders and a defensive midfield player as substitutes, but it didn't make sense to me. He said it was in case he got injuries at the back. But if you are winning, you are not going to change the defence, are you? And if you are losing, you don't put a defender on to change the game.

When he announced the line-up, I got up to go. I couldn't bear to stay and watch the game, which they lost 2–1. He told me to come back but I refused and headed straight to the nearest book-maker's. I was seeking some comfort in a familiar environment, I suppose, though it didn't bring me much. I only had about £80 and did the lot in five minutes. It was just a knee-jerk reaction. On the way through all the crowds milling in the streets outside the ground, I passed a few of the Cambridge directors but didn't say a word to them. They must have been wondering what was going on. I then caught a Piccadilly line train at Arsenal Tube station, going against the flow of the 42,000, changed for Waterloo, and took a train back to my parents in Portsmouth, having saved enough from the bookie's for my fare, remarkably for me.

There I stayed for a few days, thinking my career at the club was over. An angry Beck rang me to tell me to come back as I would be needed for the next game. Though reluctant, I decided that I ought to if I was going to get paid and not be in breach of contract. When I got back early the next week, nothing was said, though we both knew where we stood. In positions of mutual distrust and dislike. I was substitute for a game at Leyton Orient on the next Tuesday and he stuck me on, yet again, when things were not going well and the game was goalless. I promptly scored to set us on the road to a 3–0 win.

All I got when I returned to the dressing room was a haranguing for daring a backheel in my own half, however. 'Don't you ever do that again,' he said. There was not a word of congratulation about the goal, which had been a solo effort from the halfway line. Sometimes he would have a go at me if I had been in a corner, looked up and, when I saw no one in the penalty box, turned to lay it off instead of crossing it. Someone else would have got the bollocking for not being in the box if I had got the cross in, I was told. But football is not about laying blame on team-mates. It is about using your skill and judgement in a situation, taking responsibility by doing things as you see them rather than out of habit. That's what you really get paid for.

We slipped to 7th in the table with an indifferent run after the Cup hangover and in April Beck, in desperation I'm sure, named me in the team for the first time since November. I scored both goals as we beat Shrewsbury 2–1 away and moved up to 4th. In the next game, against Stoke, I got the first in a 3–0 win and followed it with another two in a 4–0 win over Bournemouth. Now we were hot, in 3rd place, and he could hardly drop me. We wobbled against Southend, rivals at the top, and Rotherham, taking only one point, but after beating Bradford City 2–1, promotion was on the cards with lowly Swansea the visitors to the Abbey in our last match. My early goal seemed to settle our nerves and, in front of more than 9000, we went on to win 2–0 to pip Southend, who were losing at Brentford, for the Championship.

I suppose I should have been elated, and in the heat of it all that day I was. But when I thought about it, it all felt so hollow – having been sub for most of the season (I was rotated with John Taylor late on, with Dion Dublin always a first choice).

Sometimes I was asked to play wide. People say that the modern game is a squad game, what with so many injuries and suspensions, but no player really feels part of it unless he is a first-choice or central to events. You might as well be the club secretary. It's all about how the fans see you. You want them chanting your name. And they weren't chanting mine very much.

It was also all achieved with football I hated, football, if you could call it that, which made Wimbledon look sophisticated. The next season Glenn Hoddle, then manager of Swindon, would have some strong things to say about us: 'We have to match them for effort and intimidate them with the ball,' Hoddle proclaimed. 'We all want success but it has got to be achieved in the proper manner. You have to play to the strength of your resources but there is a danger of raping the game.' I could only agree with him. Personally, I felt I had done well, scoring 12 goals from 16 starts and 14 appearances as a sub, but it still rankled that I had played no part in the Cup run and been abused so badly. At least, I felt, I had some principles left.

And at least I enjoyed the end of season trip to Magaluf. Because I didn't want my girlfriend Mandy, now my wife, to think I was having too good a time and wanted to get away on my own for a while, I told her the club was going to Scotland. Unfortunately she came with my dad to the airport to pick me up, to find out that the only plane due in was not from Scotland but from Magaluf. The ear-bashing I got was almost as bad as the one I received a week later when she read in the papers that Magaluf was now considered the Aids capital of the world and couples could be seen copulating on the beach.

I suppose I should have learned my lesson the following year, incidentally, when we went to Tenerife and I tried the same story. I thought I had got away with it but the trouble was, on a day trip to Wookey Hole I took her on later that summer, I met, by an unhappy coincidence, a player from Grimsby, also out there at the time, who reminded me of our drinking session together until 4 a.m. in Tenerife. Mandy was definitely sceptical about my protests of mistaken identity.

Beck would no doubt argue that the ends justified the means, that Cambridge, a small club, were now in the Second Division as a result of his managerial methods and beliefs. My argument was that we had got there in spite of them and because of a really talented group of players. Liam Daish, Phil Chapple, Alan Kimble, Chris Leadbitter, Dion Dublin and Lee Philpott were among the best players in the lower divisions, and it was as good a team as I have ever played in. No other team could have achieved what they did with the things that were being imposed on them.

I had hoped that my goalscoring record and the fact that I finished the season in a Championship team would count at the beginning of the next season, but I found myself left out of the opening game at Grimsby, which we won 4–3. Then I was back as sub for the next three matches, scoring in a 3–1 win over Watford as we burst up to second place in an explosive start.

Perhaps Beck had heard about my escapades during a one-week pre-season training camp with the Army at Blandford in Dorset, which he had organized. At the time I thought I had got away with them. We were all supposed to be resident in the billets there, but with Blandford being fairly near to Portsmouth, I thought I would take the opportunity to go home every night for more comfortable bed and board. The problem was we used to have to get up at six every morning, which meant making a ridiculously early start from Pompey.

One morning I was still half asleep and, seeing a garage,

decided to turn in on what I thought was a slip road. Unfortunately it turned out to be a grass verge and I crashed the car. I then got someone from the garage to tow it back to Portsmouth, rang my long-suffering dad and dragged him from his bed to drive me back to Blandford. By now it was getting late, or early depending on how you look at it, and it was a close-run thing as I climbed through a window that one of the lads had left open and slipped under the blanket just as the sergeant major or whoever arrived to bellow in our ears. I was all right. I was wide awake, though I suffered for it during training that day.

It meant that I had to find another car to get back to Portsmouth that night and I made the mistake of taking our goalkeeper John Vaughan's without his permission. Well, I couldn't find him, and the keys were just lying about. Now John was one of the strongest men I have ever met, built like the judo fighter Brian Jacks, and a bit of a legend at Cambridge. 'Johnny Vaughan is hung like a donkey,' a certain section of the Abbey crowd used to sing, though I don't know where they got their information from. I have waited in vain my whole career for fans to sing something as complimentary to me. John did not use physical attributes to extract his revenge, however. When I got back to Blandford the next morning, all my trousers had been cut up.

Though I have never been one for excessive physical preparation, to put it charitably, these training camps were something I agreed with Beck about. Getting up at six, doing some running, eating a hearty breakfast before more training, morning and afternoon, really did get us going for the start of the season. He was the first manager to show me that I needed to live well, to eat, drink and sleep properly, in order to be a fit, effective professional. Mind you, my gambling at the time meant I didn't always take much notice. It also built togetherness and a team spirit that I have never known at any other club despite all the reservations about the way we played. The downside was still that the 20 per cent of the game that really mattered to me, ingenuity around the penalty box, was undervalued so much of the time. That, and having to endure Beck's supposedly 'inspirational' poems that left everyone looking at each other bemused when he read them out. Is that what they mean by blank verse?

I firmly believe that this team would have made the Premier League, which was starting the next season, if only Beck had compromised a little. It was just so talented. After my early season subs' appearances, I got back in the side. That was another thing with Beck, you could have all sorts of rows with him but if he needed you, thought you could do something for him, he would overlook them and stick you in. I suppose it's what they call pragmatic, the sort of a decision a manager has to take, to be fair to him. To me, sometimes, it was more like getting him out of a hole.

Once in, I made it very difficult for him to leave me out. I got two goals in a 5–1 win over Leicester, and also scored in wins over Tranmere and Barnsley, then got the goal in a 1–1 draw at Newcastle, which seemed like a really good result to us then, even if they were struggling pre-Kevin Keegan, and showed how far the club had come. More than 20,000 saw us win a big game at Ipswich 2–1, with me scoring the winner. We were top of the table at Christmas having gone ten games unbeaten, though beginning to draw more than we won, and I suspected the new Premier League mandarins were getting a bit worried that another small club like Wimbledon were going to gatecrash their smart party. The Dons were the Crazy Gang – though I reckon Aldershot would have run them close – but Cambridge were the Bang Gang.

We were a very professional and highly intimidating side to play against. Beck had worked on things defensively in training – positions at set pieces, playing the offside trap, squeezing up on the opposition – until they were done right, to the point where the defence and midfield just took them as second nature. Also, you could see teams come out of the visitors' dressing room at the Abbey with fear in their eyes. It was a tight little ground, with not much atmosphere, being stuck out bleakly on the edge of town next to a patch of allotments. We were now playing some pretty big clubs, like Wolves, Derby and Sunderland, who just didn't fancy it. They knew we were going to bully them around.

Unfortunately I damaged ankle ligaments in November and missed the next three months. Now I am not saying that it was down to my absence but I was an integral part of the side by now and we did suffer a dip in form during that period. The real reason that the wins were drying up was that we had become too predictable and were getting found out. Teams we were beating in the first half of the season were getting wise in the second and were determined not to get caught out again. We were now no longer up against Third and Fourth Division opponents who didn't have the nous to work you out. We were now playing bigger clubs who studied videos of all your games and worked out ways to combat you, having a better class of player able to do so.

Our FA Cup run in January was a perfect example. In the third round we got a draw at Coventry and brought them back to the Abbey, an experience which seemed to scare the pants off them and they were dumped 1–0. In the next round Swindon were the visitors. We had beaten Glenn Hoddle's neat passing team 3–2 in the second match of the season, but now he had instilled in them the need to match our commitment. Then the skills and superior passing of his team took over. We were just as skilful a team, mind. It was just that we weren't being allowed to express it. We lost 3–0.

Some of us pleaded with Beck to change things, to show some flexibility, but he stubbornly refused. A home game against Newcastle was a case in point. Their centre-half wasn't even bothering to mark me. He just dropped off 20 yards because he knew I wasn't allowed to control the ball but instead had to flick it on into the corners. There, their full-backs were waiting for the ball and could control it before playing the ball upfield. Our policy of just trying to win throw-ins and corners was completely nullified and we were 2–0 down at half-time. We then asked to be given some freedom to play as Beck's way wasn't working but he told us to stick to it. I wanted to just tell him to stick it.

In fact I did later that month of March. We were just about clinging on in the promotion race, then 3rd, when a classy Ipswich team, second themselves and on their way towards winning the Championship, came to the Abbey. After about a quarter of an hour I committed the cardinal sin. Yes, that one of cutting inside with the ball and laying it off rather than playing it down the line. I was duly substituted. At half-time I was in the medical room when he came in, obviously ready for a showdown, and told me to get in the dressing room. 'I'm not doing an effing thing you say any more,' I told him. 'Just stick it up your arse.' He flew at me, trying to head-butt me, I think, but I pushed him away. He came at me again, swinging his fists, but I connected first and punched him in the eye, sending him falling back into the heat treatment equipment. It sounds quite comical, now I look back at it.

Anyway, he then took a run at me but I gripped him in a headlock and started punching him. I think all the resentments of two years, on both sides, were coming out. He was kicking and punching back at me. His assistant, Gary Peters, who had been a player with Wimbledon, came round the blindside and I was also trying to fend him off. It was pandemonium. At this point Liam Daish arrived and grabbed Beck, while Peters grabbed me and held us apart. 'I'll see you after the game,' Beck spat at me.

The game was drawn 1–1; the dropping of two points rather than our half-time interlude appeared to occupy Beck's mind when we all got back into the dressing room at the final whistle. In fact, amazingly, he was very good about it all and handled the situation really well. It did surprise me but then in football people lose their tempers all the time and say and do silly things, before getting on with life as if nothing has happened. It is a spontaneous world, which is something I love about it. The real men's men do not bear grudges.

He pointed at me as he shouted at the others. 'It's a shame you bastards didn't show the same passion he did at half-time,' he screamed. I just stood there with my mouth open. He was a bigger man in that one moment than I had ever seen him before. Perhaps I should have done it two years earlier and he might have had more respect for me. What I enjoyed most about it was going home, watching the highlights on television and seeing pictures of him in the dug-out in the first half looking fine, then in the second nursing a black eye. All my own work, I thought proudly.

It was strange for a while after that, almost (but not quite) as if all the incidents in the past had been forgotten. He was as good as gold to me. He picked me regularly in the run-in – having sold John Taylor to Bristol Rovers he had to – but we were fading due to his unwillingness to change. Shortly after Ipswich we were playing Portsmouth and I was up against Guy

Butters. When I got the ball on the touchline one time, he just jockeyed me and said: 'Go on. Go inside with it. You can't, can you?' He was right, I wasn't allowed to. I had to go down the line, as he knew I would. Then he put me and the ball in Row F. It was obvious to all but John Beck. The better players would just read the game and play out. In the lower divisions they just hoofed the ball into touch, playing into our hands, because they weren't capable of turning. But the higher we got, the more they would come out with the ball and take the piss out of you.

I scored in that match against Pompey, one we really needed to win to stay in the hunt, but we only drew 2–2. I was also on target in the next, a 2–1 defeat at Leicester, which all but scuppered us. After that, Beck came into the dressing room for another one of his psychodrama groups.

He had been hearing the rumblings of discontent within the ranks about the way we played and demanded to know who was with him and who was against him. 'I'll resign if any of you don't believe in me,' he said. 'Who doesn't believe in the system?' There was a silence before anyone dared speak up. Then it disappointed me that so few of us did, because in private nearly all the lads would slaughter the system. Only me, Colin Bailie, Tony Dennis and Chris Leadbitter said anything. It was a bit embarrassing. I think we all knew that there was no way he was going to resign. It was just a bit of kidology, him trying to see who was with him and who wasn't.

I spoke up because I believe in always being honest, that in the end it is for the best and those that do are the most to be respected, and because, after the Arsenal and Ipswich incidents, I knew my days at the club were numbered. I had made up my mind just to see out the remainder of my contract that season then move on.

By the time of a 4–2 win over Port Vale in our penultimate game, in which I scored a penalty, our hopes of automatic promotion had gone. We were then fortunate to scrape into the play-offs in 5th place with a 2–2 draw at Sunderland in the last game, one of my rare memorable days at Cambridge with me getting both goals – one a near-post header, the other smashed in after a corner – as our main rivals Charlton Athletic lost 1–0 to Bristol Rovers. We soon sensed our number might be up when we only drew 1–1 with Leicester at home in the post-season tournament, a suspicion confirmed when we were thrashed 5–0 in the away leg. A team we had beaten easily early in the season had well and truly sussed us out.

It marked the break-up of what could have been a sensational team. My contract was up and although they offered me another one, it was a half-hearted effort. Beck wasn't fussed, I could see that. Although things improved belatedly after our punch-up, there was just too much bad blood between us. My record that season was good – 12 goals from 25 starts and 4 sub appearances, including 5 goals in the last 4 games of the regular season – but I knew I needed a fresh start, the more so because my gambling was becoming a big problem at that time and not only did I need a change of environment but also a move for better money.

At an end-of-season night out Gary Peters told me he was sorry I would be leaving, which struck me as strange after everything I had been through. But, of course, being the assistant manager we understood that he would be the manager's eyes and ears whilst providing a shoulder to cry on when needed. That's the role of 'the buffer', but I didn't see it had helped me with Beck.

That night he added that my leaving would be my loss, because the quality of their system would get them into the Premier League the next season. 'You still don't get it,' I said. 'It's players not systems that get you up.' This team will get us up, he insisted. I told him that not many of this team wanted to stay. I'm not sure he believed what he was saying himself. Beck took him to Preston North End when he later became manager there, but his methods failed and he was sacked, Peters taking over and apparently getting them to play some decent football. Beck moved on to Lincoln, who in the 1995–96 season had the second worst disciplinary record in the Third Division. I noticed that in a friendly before the 1996–97 season Lincoln deliberately played with ten men for part of it to practise for those times when they would be forced to do so in the league because of a red card.

Anyway, my point was proved at Cambridge. That summer, myself, Dion Dublin (to go to Manchester United) and Andy Fensome also left the club while Colin Bailie quit the game in disgust, disillusioned with Beck's regime, to become a policeman up in Cumbria, I think. Cambridge lost their first four games of the following season, slumped to the bottom of the table, and by November Beck and Peters were sacked.

It was such a shame. But for one man's inflexibility, Cambridge could really have made history. Beck will always say that the club got so far because of him and his methods; I maintain it was in spite of them. I believe he stopped us being what we really could have become, though I do give him some credit for his ability to spot players with talent and buy them at a reasonable price. But it was no longer my problem, or so I thought. I was off to another club. Little did I know that Cambridge and I were far from finished.

6 The Abbey Habit

That summer of 1992, Steve signed for Luton Town, then managed by David Pleat. Four years on, I spoke to Pleat, now the manager of Sheffield Wednesday, to try and discover why the pair's association lasted only four unhappy months when their views on the way the game should be played coincided. 'Don't crucify him in your book, will you?' Pleat pleaded. 'He's a happy-go-lucky character and you can't help but like him. He's as silly as Mr Simple and as happy as Larry. He doesn't bear any malice towards people.' The vast majority of people who have come into contact with him appear to feel amusement and affection when the name of Steve Claridge is mentioned. Pleat, one of football's most intelligent thinkers and a manager who actually likes footballers (which may be rarer than one might imagine) feels almost protective towards him.

Pleat is also honest and realistic. 'Steve didn't appreciate other players around him as well as he might have sometimes,' is his analysis. 'He didn't hit it off up front with Phillip Gray. He didn't get his head up quickly enough, would often take one touch too many and wouldn't get the early cross in. But he was a real grafter and he would often get you out of trouble if you hit the ball up to him. He had good lungs, good legs and a big heart. Really determined, with an inner drive. He was not frustrated easily out on the field.'

Claridge and Pleat met to agree the signing at the Royal Bath Hotel in Bournemouth, where the manager was on holiday. 'The subject of his gambling came up. I told him I had heard about it,' says Pleat. 'He told me a bit about it and I found it incredible. He also said that it was behind him. But we had some sensible gamblers at Luton at that time and I got the feeling from them that he was still having a problem with it during his time with us. But I never went chasing players. It was the same with the timekeeping. He wasn't the best but I am not one of those who fines them if they are a little bit late.

'It didn't worry me that he was a bit eccentric either. He was never a roguish sort. He just got himself into wrong situations, like if he had a puncture and the tyre needed to be changed he would just leave the car by the side of the road. He wasn't that serious about life. For all that he was a lovely lad. It just didn't work out for him with us.'

It should have done. Pleat's reputation in the game was for sending out fluent pass-and-move teams of the sort Steve gazed on with envy during his thirty-month sentence with Cambridge United. During his first spell at Luton in the mid-eighties Pleat produced one of the sweetest teams in the old First Division, featuring such easy-on-the-eye talents as Ricky Hill, Paul Walsh, Brian Stein and David Moss rotating in attack like a slick basketball offence. 'When I get depressed about the game, I put on videos of that team,' he once told me. After his departure, the team would go on to win the League Cup, then sponsored by Littlewoods, under Ray Harford. Pleat himself, attracted by a bigger club, indeed one of the biggest, led Tottenham Hotspur to 3rd place in the First Division and to the FA Cup Final of 1987. The way he employed Glenn Hoddle and Chris Waddle in a five-man midfield behind a single striker in Clive Allen, who scored 47 goals that season, was truly innovative. Great things looked possible for him and the club before a tabloid got to his private life.

But any great things that might have been possible for Claridge at Luton were soon submerged. He played only 17 league games for them, scoring just 2 goals and by his own admission played the worst football of his career. His Cambridge diet had left him starved to the point where this new intricacy left him unable to adjust. He was also unused to such a *laissez-faire* regime and attitude among the players, having enjoyed, as his redemption at the Abbey, the wholehearted attitudes of his colleagues.

When Cambridge United came back to Luton to re-sign him in November, he was only too willing to go, Luton in turn grateful to get their money back and a little bit more. John Beck had been sacked after a bad start to the season, Gary Johnson taking over as caretaker manager, and on the 20th of that month, Steve agreed to a deal he would not otherwise have contemplated. He remained fond of the club and most of his old friends were still there. The one barrier had been removed.

One of the old guard was Phil Chapple. 'Our really good team was beginning to break up, with John Taylor and Dion Dublin having gone, but Steve still came back to a decent outfit,' he says. 'My overriding memory of that time is that it was a happy club, not one in decline, though I was injured a lot that year and didn't see the despondency after away defeats, travelling back with the lads in the coach.'

A problem arose within six weeks of Claridge's arrival, Gary Johnson being demoted again to allow Ian Atkins, whom Claridge came to like and whose coaching methods he enjoyed, to take over. 'I don't think Gary gave Ian all the help he might have, which was probably understandable seeing that his nose had been put out of joint,' says Chapple. 'But I think most of the lads liked Ian and appreciated his training methods. He was a players' man.

'Gary had started the change away from the long-ball football and Ian continued it. He was a very upbeat character and gave you a lot of confidence. Training did get a bit boring under Becky, with a lot of standing around. Under Ian it was in short, sharp bursts and more enjoyable. Perhaps Steve didn't enjoy it as much as I did because Ian would work with the defenders mainly and his coach Brian Owen would work with the forwards, and I'm not sure if they felt he was up to it.'

After a start under Beck that had seen the first four games of the campaign end in defeats, results under Atkins improved slightly but not enough; by spring Cambridge were still in the relegation mire. 'I think Ian made a few mistakes at first in trying to change all the things Becky had done, and only started to rely on the team's strengths, its fighting capabilities, towards the end, when we absolutely battered Southend and won away to Wolves but finished about one win short of avoiding relegation,' Chapple said.

'Often the team would play with only one up front away from home and five in midfield. The one would be Steve and you knew you could rely on him to work hard up there and hold up the ball. The theory was that the midfield players would get up in support but it didn't always happen that way. We still had a strong defence but there was a lot of pressure coming back on them.'

Ultimately, it all came down to the final match of the season at West Ham, with Cambridge needing at least a point to have a chance of avoiding the drop. At 1–0 to West Ham Cambridge had a goal disallowed, Kevin Bartlett standing offside but not interfering with play as Chris Leadbitter's shot hit the net. 'It would certainly have been allowed to stand these days,' says Chapple. 'Then they scored their second and there was a pitch invasion. I don't remember feeling frightened,' says Chapple, 'just angry because a linesman had his flag up for offside.' After going off the field for a short time, the players returned when order was restored to play out the last couple of minutes but time was up for Cambridge. They were back in the old Third Division, now the Second with the formation of the Premier League. 'At least we didn't go down without a fight,' Chapple believes. 'We went down with our heads high.'

It signalled, though, the end of an era and the break-up of what had been a potentially explosive team. Chapple would go that summer, along with several more, including the manager, who would be replaced by Gary Johnson. 'There were rumours of backstabbing and of Ian having ruffled the feathers of the directors, but I just don't know,' says Chapple. But then managers are easily sacked and cheaper alternatives put in place. More simply, the club could no longer afford the wage bill in a lower division and were keen to cash in on saleable assets.

Steve Claridge was one of those and he was on top money in the bottom half of the new Endsleigh League. Halfway through the next season, 93–94, that predator in the transfer market Barry Fry, now with some dosh in his pocket at Birmingham City, would be on the prowl.

I suppose I could have been concerned that summer about being out of contract and finding another club. Football is a precarious business at the best of times, which is why you have to be hard-nosed in negotiating contracts, and here I was unemployed. But I looked on the bright side and wasn't too worried. I was 26 and my goalscoring record had been good, despite all the aggro of the previous two seasons. I expected to be in some sort of demand because there had been several enquiries about me in my latter stages with Cambridge.

One of them had been from David Pleat, who was back at Luton as manager after spells with Tottenham and Leicester. His first successful period with Luton had been in the eighties, when they played some lovely football in the top division. It was reassuring that the phone rang almost immediately the season had ended.

We arranged a meeting straight away. In those days I did my own deals, preferring not to involve an agent, and it was the first of many over the next two months as we tried to hammer out an agreement. We started off, or at least I did, on the money I was on at Cambridge, and worked up from that. I met him all over the place; at a hotel in Reading, at Kenilworth Road and, my old favourite standby, Fleet services on the M3. Sometimes it felt like a scene out of those old spy films. It occurred to me to say at one point: 'We can't go on meeting like this.'

I also wanted to keep my options open for a while, but in the end the only other club I got close to joining was Leicester City. Their then manager Brian Little, however, had to sell Tommy Wright before he could buy, and at that time Cambridge were asking ridiculous money for me, £750,000 I think. Finally I told David Pleat I was ready to sign and went down to Bournemouth, where he was on holiday – yes, football people do lead glamorous lives and get to travel the world – and we agreed on the deal.

It was a good one: a signing-on fee of £35,000, removal expenses of £10,000 and a salary of £95,000 a year for three years. My eyes lit up. I was almost trebling my wages. There was still a problem with the fee, though, Cambridge and Luton having to go to a tribunal. The previous season, David Pleat had apparently offered £300,000 for me, only to be turned down. Now the tribunal announced, to Cambridge's disgust, that I was worth £160,000. As the figure was read out at the meeting, Pleat grabbed my arm, turned to me and said: 'Come on. Let's get out of here before they change their minds. We've got a right result.'

Pre-season and a tour to Sweden was enjoyable enough, despite the night when I lost £1500 to my new team-mates, mainly Mick Harford, at the trotting racetrack. I roomed with Trevor Peake, who was a good professional, and it was a real shock to him, I think. It was a bit like *Blind Date*. I didn't know who I was going to be in with when I went up to the hotel room. As I opened the door I saw his face drop. I think he had heard the rumours that I was not the tidiest or most organized of players. Also, I have always been a person who has trouble getting off to sleep and I like to watch TV until the early hours. That can be quite an experience in Sweden, and you don't have to understand the language, either, if you know what I mean.

I had ended the previous season playing against Leicester City, who had lost in the play-off final to Kenny Dalglish's Blackburn Rovers, and also began this one against them, though only as a substitute, as we lost 2–1 at Filbert Street. After that David Pleat stuck me in the team, but I couldn't get settled and never did myself justice at the club. I suppose, after Cambridge, I should have fitted in with Pleat's desire to play passing football but at times we played too much of it. I had gone from one extreme to another. There were no real wide players getting in good crosses and everything was off the cuff. There was no real balance between the short and the long game that I had been used to and the team had little shape or pattern. One game against Grimsby was typical. We were 3–0 down at half-time. I was unable to get in the swim of any of it against the Mariners and the whole team were floundering. Mind you, the way I was playing at that time I would have struggled to get in the swim with the Liverpool team of the seventies, which seemed to be able to take any old player and make something of them.

I stress that the problem was mainly with me, rather than Luton, who are a great little club. There were though, I felt, particular problems with them at that time. I had just come from a club where people would run through a brick wall for £250 a week. No matter what you thought of the way Cambridge played, every player really gave his utmost. They were an honest and hard-working crew. Now I had come to a club where players were getting £1500 a week and asking not to play because they wanted a move. I wanted to play because it is my nature. I have never been one to hide, even when things were going badly, and I prefer to try and play my way out of a bad spell.

The other problem was that the club had got too used to losing; it didn't hurt them enough. They had been in the top division for a few years, but losing more games than they won, and just escaping relegation a couple of times before it finally caught up with them. I just couldn't get my head round this 'Oh well, there's always the next game' mentality and it used to frustrate me when team-mates took defeat in their stride as just an occupational hazard rather than something that can be avoided now and then. There were good, tough professionals there like Trevor Peake and David Preece, but others who weren't up to it, as well as some kids learning their trade. It didn't make for a very successful blend.

I did have a few good games but nowhere near enough. I scored three goals over the two legs of a Littlewoods Cup first-round tie against Plymouth Argyle but we lost 5–4 on aggregate. I also scored in a 3–3 draw against Tranmere in September, only our second point of the season, but that result saw us slip to one off the bottom as we had failed to score in our previous three matches. My only other goal was a penalty that gave us a 1–1 draw with Birmingham City.

They totalled up to the worst 17 games I think I have played. During it the local paper, the *Luton News*, carried a letter from a fan highlighting David Pleat's buys and summing each one up in one line. In horse racing parlance that might have been written so I could really grasp it one player was labelled a non-trier and I was described as a donkey. It was a fair assessment. If I haven't done well, I am the first to hold my hands up, and at that time they seemed permanently above my head.

I don't blame David Pleat. He was an excellent coach, different to anyone I have had before or since. Mostly I had been with disciplinarians but he was a more technical and tactical coach. He also tried to work you out mentally, I reckon. It's not that he plays mind games as such, more that he likes to be Mr Psychology to try and get to what makes a player tick. I think he nearly developed a nervous tic, though, trying to work me out.

No matter how promising the circumstances look sometimes, it can just be that you are with the wrong club at the wrong time, and that's how it was for me. I believe I have always been able to hold my head up in any town I have played in, because I have always given my best, but Luton was probably the place where I let myself and everybody else down. I suppose I should have seen the signs. I bought the flat in Wimbledon and stayed with Crystal Palace for just three months. After all my accommodation adventures in Cambridge, I decided to settle down near the club and I bought the house in Luton. Again, I lasted a mere three months.

In November Luton needed to sell a player to pay a tax bill on the transfer that had taken Mark Pembridge to Derby County for more than £1 million. My number was up. I don't think David Pleat had quite given up on me but when Cambridge United, who were really struggling for goals, came back in for me – I was probably offered back to them, truth to tell – he had little choice but to go along with it. After all my problems at Cambridge I would never have entertained the idea, except that by now John Beck had been sacked and Gary Johnson had taken over as caretaker manager. I had always got on well with Gary, who had been demoted to the youth team when Beck recruited Gary Peters as his assistant. And he'd also always been uneasy with Beck's methods at the club, even if he was the one who first thought of the cold showers. Since then he had kept his head down until his chance came. When it did, he recommended me to the board again as a safe bet for goals now that Dion Dublin had left the club.

Cambridge were offering £195,000 for me – £35,000 more than Luton had paid – so it was no surprise that Luton bit their hand off given my form. It is strange how your value can go up when you look to have become a worse player, but then people will pay depending on how great is their need or desperation, and Cambridge were getting themselves into a desperate state. It would not be unfair to say that some players do cause trouble or play badly in the hope of getting a lucrative move but that has never been my way, despite having needed the money at times. You soon get a reputation in football which in the end is self-defeating as managers steer clear of you. Besides which, you are cheating the public and yourself.

Anyway, no matter the reasons, Luton seemed pretty happy to be offered the money and one day I was called out of a team meeting to go and discuss terms back at the Abbey Stadium. Later, one of the Luton lads told me that Pleat had informed the meeting that I was on my way to Cambridge – though it might have been via Newmarket, he added. This time I managed to find my way to Newmarket Road without the need for a lift from a stranger, however.

I needed the move for several reasons. Foremost was the football, and a desire to end what had become a demoralizing situation at Luton. Second, the money helped. I was to get a signing-on fee of £50,000 and a salary of £110,000, more money than I had ever dreamed of and which would make me the highest earner in Cambridge's history. How they afforded it I'll never know, but I wasn't about to argue. I also needed a change of scenery – though these days I recognize that it is me that needs changing rather than the scenery. The arrangements all seemed good as well. Despite everything, I still had an affection for the club and the honesty of its players. The one man who had spoiled it all had gone. I could also stay in the house in Luton and travel to training every day, and perhaps all that time in the car would stop me frittering away the money I was going to be earning.

This time I did enjoy it. The club were clearly going to struggle to stay in the Second Division after a dreadful start under Beck, which saw them lose their first four matches and slump to the bottom, but there was enough in place to suggest they could escape. There were still some good players there like Liam Daish and Phil Chapple, forming a solid central-defensive partnership, Alan Kimble, Chris Leadbitter and Lee Philpott. Perhaps allowed to play more football, we might even recover from the damage Beck had left in his wake.

From my point of view, first impressions were encouraging. I scored in a 2–2 draw at Leicester – who seem to have become a recurring theme through my career – and we felt that maybe we had stopped the rot. The size of the task was apparent in the next game, though. Although I scored, we were soundly beaten 4–1 at Newcastle, who were now flying under Kevin Keegan's management and would go up as champions. You could feel then the undercurrent of enthusiasm that is now evident at St James' Park.

Soon Gary Johnson was demoted again and Ian Atkins came in from Birmingham City to be player-manager. Ian was good, and I quite liked him. I thought he had some sensible things to say and he laid on good training sessions – when he took them, that was. The problem was that he didn't do it frequently enough, preferring to leave it to his coach Brian Owen, who was really a physio, and too often we weren't organized, which was a change for the worse after John Beck. If Ian hadn't delegated as much responsibility to Brian and had more day-to-day input himself, I think we would have benefited.

Steve Butler was signed from Watford to give me some support up front, and all winter we battled away gamely. It was a struggle but never did it seem as depressing as during the Beck era, even though our results were mediocre. A run to the fifth round of the Littlewoods Cup, for which I was ineligible having played for Luton, gave some cheer to the club, but when it ended with a 3–2 defeat at Blackburn we were back in the mire of a relegation battle. There was the odd highlight, a goal here and there, but not enough. I was the club's top scorer with 7 goals, which tells you everything.

We did give ourselves a chance of staying up in our penultimate match when we earned our best victory of the season, 3–1 at home to Southend, with me getting the first goal. At that time, the decisive factor for teams level on points was goals scored rather than goal difference, and Sunderland, above us, had one more point and had scored one more goal than us. It meant that if we won our last game while they lost, we would stay up. We could also pip them if we got a point from our last game, scoring at least one goal in the process, while they lost. They had to go to Notts County. The problem was we had to go to West Ham, who were desperately fighting for promotion themselves. They were level on points with Portsmouth, their rivals for the second automatic promotion spot, but had scored a goal more. Thus they had to match, at least, Pompey's result against Grimsby that day. As a Pompey boy, it was built up into a big occasion for me.

It was all complicated stuff as it often can be on the last day of the season given all the permutations, but simplicity was really the order. West Ham had to win and we had to win to be sure of achieving our respective goals. Something had to give.

Upton Park was high on energy and anxiety, with 27,399 fans turning it into a den of hostility. We kept our nerve well enough though, and held them scoreless at half-time. Then David Speedie gave them the lead in the second half and West Ham thought they were up, with Pompey 1–0 down at home to Grimsby. Word came through to us, meanwhile, that Sunderland were losing at Notts County. A goal and a point could still do it for us.

Then Chris Leadbitter hit a tremendous shot into the net and we thought we had done it. Just as we began to celebrate, though, we saw a linesman's flag raised. Kevin Bartlett was adjudged to have strayed offside but there was no way he was interfering with play. In these days of 'passive offside', the new regulations decreeing that the benefit of doubt should go to the attacker, it would surely have stood. But now we sensed our chance was ebbing away. The home fans were delirious, thinking that they were already promoted, but with Pompey having turned their match around to lead 2–1, the Hammers actually needed another goal to be sure. There seemed to be 15,000 people on the pitch when they got it right at the death. It might have been one of them who scored it for all we knew, though it was credited to Clive Allen. They were up and we were down. That second goal was probably for the best. We may not have got out of there alive if it hadn't gone in.

At the final whistle I was as frightened as I have been on

a football field, apart from that time with Weymouth against Wealdstone, though I have suffered worse. In fact I have been struck by spectators three times. The first was when I was with Bournemouth, playing a match at York. I was going about my business in retrieving the ball for a corner behind the York goal when I suddenly felt this fist around my face coming out of the crowd. I was just stunned and stood there looking around for a moment or two. It was impossible to identify the culprit, as the coward had slipped back into the crowd. He would not have made a professional boxer, anyway, I know that.

The next time was during my first spell with Cambridge United in a match at Birmingham, which was to give me a taste of, shall we say, the passion of the home fans, which I experienced in full later. We were 3–0 up and they were fuming. Fans were running on to the pitch during the game, making for the grandstand where their then manager Dave Mackay was sitting and tearing up their season tickets in disgust on the touchline in front of him. At the end of the match, as I was trooping off, I felt a slap around my face from behind and turned to see a dirty pair of heels disappearing into the distance. I just let the incident go.

The other time was when I was with Birmingham and playing at Stoke. I remember it vividly because I was a substitute and after I came on with us 1–0 up I was helping out in defence at one of their corners. The ball was going out for a goalkick when I leapt like a salmon right on the byline, having lost my bearings, kept it in and diverted it into the path of Mickey Thomas, who, thankfully, volleyed it over the bar. It did me some good, mind. After that I was told not to come back in defence but to stay upfield. Moral of the story: cock up a job and you get out of it. No, not really.

Anyway, earlier, when I had been sitting on the subs' bench,

I had been watching the game when a fist from behind the dug-out stretched out and clipped me round the side of the head. It disappeared as quickly as it appeared. I turned round and there was no chance of finding the bloke, let alone reporting him. I have always kept a wary eye out after that, preferring not to get too close to the crowd. It was also the reason why, when Leicester won at Stoke in the play-offs later in my career, I was off the field and into the dressing room as quickly as my legs could carry me, telling the other lads to do the same.

I have never been tempted to retaliate, though what Eric Cantona did at Selhurst Park is understandable given some of the crap we have to put up with from a wide variety of abusive nutters. Most of it goes unnoticed, except in high-profile cases like Cantona's, and every pro probably has a story of being struck. Any response is just not worth it, though, as Cantona found out.

If you rise to the bait, it is only going to get worse. It is best to take it in good spirit at best, quietly at worst. It is an unavoidable fact of the people's game that it inevitably attracts some pretty strange people, and when they pay good money to watch you, they are going to want to have their say. You are always going to get a bit of stick; in fact, it shows that the crowd have noticed you and that away supporters think you are a bit of a threat. I try to smile if I can, which has defused several situations. People have told me it is one of my better attributes.

Our relegation back to the Third Division cost Ian Atkins his job, after less than a year, which enabled Gary Johnson at last to get the position on a full-time basis, as many of us felt he deserved. Ian actually thought I had engineered his dismissal, perhaps believing that as the club's highest-paid player I had some influence. He rang me about this and I assured him that this was never the case.

Gary's problem was that he now had to sell players as the club revised their budgets downwards. There were plenty of takers, which showed how highly our players were thought of and confirmed my previous opinion that it could have been a class side had it been allowed to fulfil its potential two seasons earlier. But Lee Philpott went to Leicester, Alan Kimble to Wimbledon and Phil Chapple to Charlton, all bettering themselves. In December Liam Daish would also go to Birmingham City. The only surprise to me was that Chris Leadbitter could do no better than Bournemouth. He really was a classy midfield player with a sweet left foot.

Not surprisingly we struggled at first, despite a 3–2 win over Blackpool in our first game, and took only a point from our next four games. The League Cup, now sponsored by Coca-Cola, provided us with some respite and paired us, wouldn't you know it, with Luton. We won both legs 1–0 and I scored both goals to keep up my developing record of always finding the mark against my previous clubs. Perhaps I have had a lot of points to prove. Then again, I have also had a lot of previous clubs. In the first leg it was a simple sidefoot into the net after the ball dropped to me, in the second a header at the far post. David Pleat caught up with me after the match. 'You couldn't do that when you were playing for me, could you?' he cursed. Any dreams of glory were short-lived, though. We lost in the next round to Ipswich, 4–1 on aggregate, me getting the lone goal.

For a while we improved in the league as we got used to our new straitened circumstances and Gary Johnson's more benevolent regime took shape. I was averaging a goal a game as we settled into mid-table, a fair reflection of a fair-to-middling side. The reduced ambitions of the club – I think they call it downsizing in business these days – were obvious, though. Under Beck, for all his faults, the club had been well organized and there had been money to spend. We were always well kitted out and well looked after in the peripheral things like meals after training and overnight stops. Now corners were being cut. Some of the kit was scruffy, no longer were we fed after training – though I often used to miss the meals anyway in my rush to get to the bookie's during my first spell – and the hotels, when we did stay overnight, were of poorer quality.

Still, I was quite enjoying it. But these were quiet times the like of which I had never really known. I felt like one of those soldiers in the trenches: 'I don't like it, Carruthers, it's too damned quiet.' I knew something must be about to happen. It always had in my career so far.

Word reached me that my old Fleet services acquaintance Barry Fry had been to watch me in a match against Exeter City over Christmas and afterwards had met with the Cambridge board to discuss a deal. Barry had just gone to Birmingham from Southend United amid much acrimony and was obviously looking for a centre-forward or three. He always did have an amazing memory for players, as he was out watching games every night at all levels. It was his main strength.

Now it seemed he had money to spend and he was not going to waste the opportunity. I reckoned he was being shrewd going to the board, rather than Gary Johnson, in trying to buy me, which was illustrated when the board agreed to let me go for £150,000. It showed how desperate they were to reduce the wage bill, to which I was the chief contributor, and how little they knew about the game.

Gary was angry when he heard. Although he knew he would have to sell me, he was determined to get the right price for the club, even if they didn't deserve it. In the end the fee was set at £350,000.

Barry didn't really have to sell Birmingham to me when I arrived at St Andrews. Having played there a few times I knew this was a real football centre, full of passion for the game, and a place where I could thrive. I know a lot of clubs get the 'sleeping giant' tag attached to them, but with Birmingham City this really was the case.

The first face I saw when I arrived in the reception area was that of Liam Daish, who had just left Cambridge as well and was waiting to see Barry about his contract. In fact a load of us were. There seemed to be a conveyor belt into his office, with myself, Paul Harding, Steve McGavin and Roger Willis waiting to sign. It was a bit like a doctor's surgery, with everyone getting their ten minutes then being told to go away and think about it. During my ten minutes, Barry and I recalled our meeting at Fleet services when he was manager of Barnet. 'I've got more money this time,' he said. 'But you'll have to pay all the tax now.'

I didn't realize then that this would just be a normal week's work for Barry and would go on for another two years. When we'd finally agreed terms, some time later, we were all unveiled at a press conference, along with a few others. In fact, I think there were more players than press there.

The time I signed for Luton it took me two months to agree terms with David Pleat so at Birmingham, for the first time, I used an agent – Johnny Mack from London – at a cost of £1500. I still didn't get the deal I wanted and, in fact, I took quite a pay cut to sign; the £90,000 a year for two and a half years being £20,000 less than at Cambridge, though I was to get a signing-on fee of £30,000 a year. My understanding of the negotiations, when I was told that the situation would be reviewed at the end of the season, was that this would mean a pay rise. This was a massive club that seemed to be going places and I wanted to be part of it.

Though I thought the money was a bit stingy, I was soon to find out that it was still too much for some at Birmingham, including their new managing director Karren Brady, former first lady of the *Daily Sport* newspaper and the owner David Sullivan's right-hand woman, a duo at the head of the club who were attracting plenty of the publicity they craved. Soon, too, there would be Birmingham fans telling me what a waste of money I was.

7 Rebirth of the Blues

Birmingham is a city divided by football. Around its boundaries
other large clubs co-exist, and games between them are
more or less local derbies; Wolverhampton Wanderers and
West Bromwich Albion to the west, Coventry City to the east.
Within England's second city itself, though, you are either the
claret and blue of Aston Villa or the blue and white of
Birmingham City.

Villa are the aristocrats. Their stadium, tucked between the
M6 and M38 Aston Expressway, is a monument to the
prosperity of the Premiership, in whose top tier they are
established. Birmingham, their St Andrews ground located
amid railway lines, industrial estates, wasteland and the
functional housing of the working-class suburb of Bordesley,
have existed in Villa's shadow for virtually all of their own
121-year history. Their supporters, noisy and intense, are

infused with the resentments of poor relations, their underachieving club having let them down time after time. Best season: 6th in the old First Division in 1956; losing FA Cup Finalists that year and 1931; winners of the fledgling, then barely regarded, League Cup in 1963. But still the fans continue to turn out in numbers ready for the glorious new dawn. The place and the people are passionate and earthy. It was to be an appropriate home for Steve Claridge.

In 1993, Birmingham were acquired by David Sullivan, an Essex publisher of soft pornography whose stable, or should that be kennel, of publications included the more down than downmarket tabloids, the *Daily* and *Sunday Sport*. He installed Karren Brady as his managing director, who had been responsible for the marketing of the *Sport* publications. City fans welcomed the change. Under the previous owners, the Kumar brothers, producers of bargain clothing, St Andrews had become a graveyard of ambition. The club's song urged them to 'Keep Right On To The End Of The Road' and at times they had seemed near it. Underinvestment had seen the ground rust and the team mostly languish in the Third Division, though now they were languishing in what had become, with the Premier League's inauguration, the First.

Sullivan gave the incumbent manager Terry Cooper a chance, but when a relegation campaign seemed to be developing again in the 1993–94 season, he was dismissed. It seemed almost kind after his recent experiences: for you, my son, the struggle is over. In, poached from Southend United – for which the Football League would fine Birmingham heavily – came Barry Fry, the chirpy former manager of Barnet, one of the game's most ebullient and quotable characters. It was to be the start of a remarkable period for the club and the English game, with Sullivan, Brady and Fry all grabbing headlines and attention, the events at the club worthy of a soap opera.

Finally given some spending power, Fry set about accumulating the players he had always coveted but had been unable to afford. Steve Claridge, his quarry on the M3 six years earlier, was one such and, on January 7, 1994, for a fee of £350,000 he was secured for Birmingham City from Cambridge United.

Over the next 27 months, Steve would see players arrive and depart at St Andrews (mostly arrive) and would play alongside myriad partners in attack. Fry, a football-daft figure who had suffered two heart attacks, had watched games four or five days a week at Barnet and Southend. He was a walking, waddling, who's who of the underbelly of the English game. I have been in his office when another manager at a Third Division club rang up to enquire if he could recommend a good centre-half and heard Fry reply that he had just seen a decent one at Braintree. Braintree?

Most of his former favourite players followed him to Brum, and all manner of others were acquired, either expensively or on loan. Birmingham City, the reception area now in need of a revolving door, could have supplied many a wedding with something old, something new, something borrowed and something Blue. Though acquiring a reputation for ruthlessness within the club, Brady, backed by Sullivan, indulged Fry for long periods.

Fry could not quite keep Birmingham in the First Division despite a spirited finish to the campaign. The following season the manager endured the pressure early on from Sullivan, being told publicly that if the team were not in the top three by November he would be sacked.

Gradually it came right, however, and Birmingham were promoted as champions, on the way lifting the Auto Windscreens Trophy at Wembley by beating Carlisle. An astonishing 50,000 of their fans were there to see it. They established several club records: most points in a season (89); longest unbeaten record of 25 games (20 in the league); and most league goals since 1967–68 (84). It was also a personal triumph for Steve. After another indifferent start with the club, he became the supporters' Player of the Year and the first Blue since the club legend Trevor Francis to score 20 league goals. It was all achieved with a mixture of the long and short ball, the attempted expansive and swift game that Fry had been nurtured on under Matt Busby as a Manchester United apprentice.

The late Sir Matt would probably have raised a few eyebrows at what went on inside St Andrews, however. It remained a crazy club. But then Steve knew all about craziness. 'Where do you start with him?' wonders his former team-mate, the central-defender Chris Whyte. 'He is one of the nicest guys I have ever met, he's just so funny. But he's daft.' Adds another colleague, the right-back Gary Poole: 'He was forever on the move, always here, always there. Worst trainer in the world. Always something on his mind.'

Whyte and Poole, also from the South, would travel up to training each day with Claridge, meeting him at Toddington services on the M1. 'He would never drive,' says Poole. 'We used to moan at him but he always said he had some problem with his car. Never gave you any petrol money.' Whyte recalls one day Claridge asking them to pull in at the petrol station at Corley services on the M6. 'We thought he wanted to go into the shop but he just disappeared round the back of the car. We sat there waiting for about ten minutes, then he came sprinting back, jumped in, said, "Come on, let's go," and offered us some chocolate.' Where had he been? 'You never knew with Steve,' says Poole. 'He could be very secretive. Perhaps he dashed back to the main services to the cashpoint for some money to put a bet on.'

They knew all about the gambling. 'I had to lend him a couple of hundred quid once to put a bet on,' says Poole. 'I got it back all right. He was as good as gold about that, although I think he still owes me two quid. One day I had four pounds on the dashboard and the next day I noticed there was only two. I asked Steve if he had had it and he said, "Oh yeah. I bought a sandwich with it." He would also ask to borrow my club car for ten minutes and bring it back a few hours later.' Then there was the singular eating style. 'Whenever we stayed in a hotel, Steve would always grab the pineapple, you know the one that's usually there for decoration, from the lobby or the restaurant to take up to his room to eat.'

Poole was one of the Fry favourites, having followed him from Barnet and Southend. 'I really liked him. There are so many good things about him. He was a good motivator and a great character, a one-off. I remember one time he was having a go at our midfield player Paul Tait, calling him a skinny bastard. Taity came back with "You fat bastard." They were standing there face to face for about five minutes, one shouting, "You skinny bastard," the other shouting, "You fat bastard."

'It was just the insecurity of never knowing if you were going to be in the team, with so many players about the place. Baz tried to make it like Wimbledon but they were more professional. I think Steve mostly got on pretty well with him as well. Except the times when he was taken off if he had done a couple of things Baz didn't like.'

Steve, though, for most of the time was one of Fry's few fixtures in the team. As defenders, both Whyte and Poole appreciated his playing qualities. 'He was a different class. I really rated him,' says Whyte. 'He runs all day, has got good touch and he'll give you something to aim for.' Adds Poole: 'I used to love playing with him. He could be down the left wing one minute, the next you would lift your head and see him on the right. He was always working across the line, always showing for you. In the local paper up there they put in a little description summing up how you had played. Steve's would always be "hard-working" or "workmanlike" or "workaholic". We used to wind him up about it and he'd bite back. "I'm not just a hard worker, I'm skilful. I'm a creator," he used to say.'

In Claridge's third season hopes were high at the newly promoted club that they could go straight through the First Division and into the Premiership. By now, the ground had been three parts rebuilt, and beautifully too. After a promising start, which saw Ipswich, Norwich and Barnsley all well beaten, their form was inconsistent, however. Steve enjoyed a purple patch in October, scoring six goals in the course of four consecutive wins, but the season was ultimately to tail off.

It was also soured by experiences in the Anglo-Italian Cup, when the match between Birmingham and Ancona in Italy turned into an ugly brawl involving both sets of players in the tunnel immediately after the match. Steve, on the bench that day, will tell of it in detail. Gary Poole recalls something altogether more bizarre from Barry Fry, who has been noted for his touchline tours down the years, particularly when his teams score and, Chairman Mao cap atop, he sets off on an air-punching run of celebration. 'I was with Steve on the bench,' Poole remembers, 'and we were just in fits of laughter. Baz used to shout things at our players and he was calling one of them a chicken. Then he started running up and down the line flapping his arms and making chicken noises.'

Partial redemption that season came in City reaching the semi-finals of the Coca-Cola Cup but they did themselves little justice against a mediocre Leeds United team in losing 5–1 on aggregate. The first leg was a personal highlight for

Steve as he was named ITV's man of the match but the second a sad experience as he missed a penalty that might have got them back in the tie.

It proved to be his last match for Birmingham City. It was also to be one of Barry Fry's last matches. The play-offs were a possibility at one point but the team faded to 15th. David Sullivan decided Fry had gone as far and as high as he was likely to and they parted company. Barry bought into Peterborough United to become a director of football and the process of buying and selling began anew.

Sullivan appointed the club's favourite son Trevor Francis as manager, a job he had performed before at Queen's Park Rangers and Sheffield Wednesday, with modest success. The owner's profile had been raised by his association with Fry and Birmingham City. Now he wanted some dignity and credibility to go with it. And, with his stadium now fit for it, Premiership football. Francis's task was to sweep away much of the quantity that had been stockpiled by Fry and bring in some quality.

For Steve, it was a lively and successful period of his career. He would come to look back on it fondly, especially his relationship with the fans, with whom he identified and who took him to their hearts. He was, while it lasted, a true Blue.

I was a bit late, as usual. Well, the traffic was bad. After Aldershot, Luton and Cambridge United, I wasn't used to such big crowds at a match. It was my first home game for Birmingham City, against Sunderland, and I was a bit unsure of the lay of the land around St Andrews – where I could park, how to get there, that sort of thing. I saw a pub near the ground called the Watering Hole, pulled into the car park, left the car there and dashed to the ground.

I had such a bad game it was untrue. After only 10 minutes, during which I don't think I had touched the ball, I went to retrieve it for a throw-in and a bloke ran 20 yards down the stand to shout at me: 'Claridge, you tosser.' My first thought was that my new club and fans were going to present an interesting challenge to look forward to. Or something to that effect. It got worse, the game was goalless and the fans went home grumbling.

Wearily I made my way back to the Watering Hole and was just putting my key in the car when I noticed a rather large gentleman emerging from the pub, a little unsteady on his feet.

He seemed to be coming towards me, perhaps to exchange analyses of the match, I thought. In fact the only thing he wanted to exchange was punches. 'Oi, Claridge. You had an effing Freddy,' he ventured. I took this to mean a nightmare, but thought it a little premature and told him so in no uncertain terms, as his mates held him back. Actually, I think I managed to blurt out: 'Um. Just give me a little more time.' Like one minute to get out of this car park. I was gone in 60 seconds.

Mind you, it was a better end of the day than the previous Saturday at Watford, my second game for the club after making my debut at Notts County. We were a shambles. Nobody knew where they were playing or what they were doing and we found ourselves 5–2 down. The whistle blew for a foul and I stopped. All of a sudden my marker David Holdsworth came through on me late. I just turned round and booted him in the leg, for which I was sent off.

Thus after the Sunderland game I started a three-match suspension. The problem was we had no midweek matches at the time and it took nearly a month to serve. Sometimes you can do it in a week if you're lucky. I also caught the flu and was really low. So too was the club. They had gone out of the FA Cup in disgrace, 2–1 at home to Kidderminster Harriers of the GM Vauxhall Conference, and were in the middle of a thirteen-game run without a win that sent them to the bottom. With the Premier League having been formed, we were now in the First Division, formerly the Second, having come up from the Third the previous season. But for all the changes, whichever way you looked at it, relegation looked certain and the Premier League, for a club which thought it should be up there, looked a long way off.

'I made a right mistake with you,' Barry said to me after I had been there just over a month, and he was ready to give up with me. He was going through a difficult time himself, what with the team struggling and Birmingham just having been fined £130,000 by the Football League for inducing him to leave Southend. I did score a couple of goals, to earn us points against Derby County and Luton, but they were slim pickings and my customary bad start with a new club was being acted out again.

What topped it all was coming home from a match with a couple of the other new signings, Richard Huxford and Chris Whyte, one Saturday night and tuning into the Birmingham radio station BRMB. A caller was going through all Barry's signings one by one and giving his assessment of them. There were so many signings it took about half an hour. I knew the bloke would get round to me eventually and I was sinking lower and lower into the back seat, cringing. The other lads escaped relatively unscathed and were laughing by now. 'And that Steve Claridge,' the caller finally said, 'I have been going to St Andrews for thirty years now, and without doubt the worst player I have seen in my time is that Steve Claridge.' Now Hux and Bear, Chris's nickname, were falling about laughing, the car weaving about on the motorway a bit. I couldn't get any lower. I just had to laugh too.

Just before the transfer deadline Barry told me he was going to sell me – I think he was getting restless as he hadn't bought or sold a player for a few days – and that I could go if he got a decent offer. In fact, after a 1–1 draw against Luton, in which I got the goal, he told Anglia Television that my transfer was just a question of who and when. Two clubs came in for me: Leicester, again, and Norwich, now managed by Martin O'Neill. But every time they made a bid, Barry would up the price and they both backed out. Baz is the sort of manager that defines your market value by who wants you. If Gillingham had come in, I could probably have gone for ten grand. If it had been A C Milan he would have asked for ten million.

So the deadline passed at the end of March without me being sold and it worked out for the best all round. I suddenly felt more settled and so did the team. Barry's *raison d'être* was taken away from him as he could no longer wheel and deal before the end of the season. In fact he might as well have gone on holiday. He had to go with the players he had got and instead of all the chopping and changing, coming and going, we fielded near enough the same team for the remainder of the season.

There may have been another factor that helped. The week before the deadline we had just lost 3–0 at home to Leicester, our sixth defeat in seven games. Barry came into the dressing room. 'You lot got anything to say?' he asked. 'Because something has got to be done. What's the problem round here?' I had been dropped to substitute and had the right hump. Although I hadn't been there very long, I felt I needed to speak out. 'You're the problem,' I said. 'You don't know what you're doing so neither does anyone else. You don't know what team you are picking and nobody knows the formation. You're just dropping people on a whim.' A couple of the others then spoke up in agreement, Liam Daish and Paul Harding. Barry actually took it well, like a man. He admired honesty and people showing their passion for the game and the team.

At this point David Sullivan came into the dressing room and said there was a bonus of £88,000 on offer to be spread around for the squad if we stayed up. I don't quite know how he got that figure but it sounded good to us, even if it did only work out at about a tenner each with the size of Baz's squad. Perhaps Sullivan felt his money was safe as we were six points adrift of the team above the bottom three and they had two games in hand. Now cynics will say that the quickest way to a footballer's heart is through his wallet, but whether it was that, the clear-the-air talk, or Barry's decision to stop tinkering with the team, we started winning and I started scoring. It began to look as if we might just scramble clear, that the proprietor might be parting with his money.

It started with a run of three home games in April. First we beat Stoke 3–1, with me scoring a goal, then Southend by the same score. The following Saturday we were 2–0 down at home to Bristol City but when I converted a penalty we were back in it and managed a draw. Unfortunately we now had four away games to finish the season but when we had an excellent 2–0 win at Portsmouth the following week, in which I scored another penalty, we really thought we could do it. From being bottom, seven points adrift, we were now just one place below the trap door.

The next Wednesday we had to go to West Bromwich Albion for what was a massive local derby with national repercussions, a match which would be watched by more than 20,000 people. Every game was make or break by this time but this was something special. Peterborough United were already down and we had by now dragged Oxford United back into the mire. Albion were fourth from bottom, we were third and if we won we would be just a point behind them with two games left.

It was one hell of a night, one of those you live for as a professional footballer. The sight of the floodlights gives you a surge of tension as the coach nears the ground, the packed stands seem to give you an extra charge of adrenalin, and the night is heavy with atmosphere and importance. It turned out to be one of the most memorable matches of my career. We started badly and found ourselves a goal down. Then Scott Hiley made a break down the left from full-back and sent in a shot that took two richochets, wrong-footed the goalkeeper and came to me a yard out to tap in. You know when you get chances like that it is meant to be your night. After that we just tore them apart and went 3–1 up before they pulled a goal back. Then, with Albion pressing for the equalizer, Scott lobbed a ball from deep over their defence, I ran on to it, rounded the keeper and sealed the game. At the end of a breathless, magnificent night, I had scored twice and we had won 4–2.

The atmosphere in the dressing room was really buoyant, not only because the £50 win bonuses were mounting up and the £88,000 was in sight. I have never really bothered about win bonuses too much, anyway. I learned early on at Aldershot not to count on them. We were also due a bonus for getting out of the bottom three but our wages were made up on a Monday and during the intervening weekend Albion won while we drew, falling back in the relegation zone, and lost out.

The next Saturday we got a really good 1–1 draw at Bolton, who'd had a fantastic FA Cup run that season putting out Everton, Arsenal and Aston Villa, and with Albion losing 3–2 at Luton we were all square on points going into the final weekend of the season, us at Tranmere, them at Portsmouth. They had scored seven more goals than us, however, so we needed to draw if they lost, win if they drew. If they won, it was pretty much all up for us, as we would have needed to have topped their victory by eight goals. Oxford needed both Albion and us to lose, and themselves to win, if they were to stay up.

Tranmere had had a good season, nursing hopes of automatic promotion for much of it until wobbling towards the end. They were assured of the play-offs, though, which offered us some hope. We reckoned that they had little real incentive to win apart from re-establishing their form for the post-season tie. It turned out that way. They were not really up for the game as much as we were. Things looked good at half-time as we were leading 1–0 and West Brom were only drawing 0–0. We played our socks off in the second half and came away with a 2–1 win but all the while we were achieving it we knew things were going wrong. We heard a groan from our huge support which was confirmed by the bench telling us that West Brom had scored at Pompey and when we heard no more noise from the crowd, guessed that our number was up.

It was a sad end to what had been an emotionally charged day, indeed, last six weeks of the season. The sadness was not as intense as it might have been, though. We were convinced that we could bounce straight back up as long as the manager didn't tinker too much. But that was a bit like asking a piranha not to bite; in my 16 games I had already had 8 different striking partners.

I also felt I had done pretty well in the run-in, scoring 6 goals in those last 10 games, and thought that the summer would be a good time to talk to Karren Brady about the pay rise I had understood they had promised me if things went well. I had accepted their offer as I was keen to join the club but had sought an assurance that I would get a new contract at the end of the season, to get back to the money I had been on at Cambridge United, plus a rise on that.

I was in the middle of telling her that here I was back in the Third Division on less money than I had been getting at Cambridge, and that my understanding from Barry was that I would be getting a pay rise if everything worked out, when she announced that she had to walk her dog. So up she got and walked this fluffy little thing around the car park to do its constitutional. I can't recall what breed it was but it looked a bit like Tin Tin's dog, or one of those that Michael Palin kept dropping pianos on by mistake in *A Fish Called Wanda*. She came back to say that it was her job, not Barry's, to determine pay rises and that I would not be getting anything on top after they had made up my money to my Cambridge salary.

I knew I had no chance that day; Karren had such a strangle-hold on the club. I felt very let down and asked to go on the transfer list as a last resort, although I didn't want to leave. I really felt Birmingham were on the verge of something and the atmosphere around the place was vibrant despite the relegation. A spanking new stand covering half the ground was going up with the club saying the stadium was going to be the Old Trafford of the Midlands and for the first time since Crystal Palace I felt I was at a big club.

I stuck to my guns, though, and went to see her time and again. I suppose persistence paid off, or perhaps they did it to shut me up and keep the supporters happy – who had by now taken to me (there was a lot of correspondence in the local press saying that Birmingham should hold on to me) – because finally they relented. One hot, sunny day pre-season we were playing at Leicester's training ground; three games involving the first team, reserves and youth. I wasn't playing because I was injured. Barry called me over at the touchline and handed me his mobile phone. 'Go on, son, do your deal, then,' he said. On the other end was David Sullivan offering me an extra £5000 a year. It took me up to £115,000 when I wanted £120,000 but it still didn't seem enough; after all, Mark Ward, who had come from Everton, was on £175,000.

At that time there were varying salaries within the club, which caused big problems with some of the lads, particularly those who had come from non-league football, initially grateful for the pittance and the chance to be a full-time pro but gradually getting resentful. At least it was a rise. I had won my matter of principle, I felt, though I was left feeling like some *Sunday Sport* employee. Football is not a normal job and can never be, given the precarious nature of the business and the personalities involved. You are in it for a short space of time and you never know what's round the corner, so you have to get what you can while you can. If it was a job guaranteed for life, you would probably take £500 a week.

That summer Barry took us to Dunstable Downs when we reported back for pre-season fitness work. 'We've got hills round Brum,' somebody said to him. 'Not like these you haven't,' he said. It was a favourite torture chamber of his from the days he used to manage the town's Southern League team. Ever the showman, he had once got George Best to play for them. I thought it would be great as I only lived ten minutes away at Luton, just over the M1. I didn't have to get up until 9.45 a.m., while all the other lads were moaning about having to set off from Brum at 8.30.

I liked the way it started, too. We would begin with a bit of ball work at the bottom of the hills. All the while, though, you were looking up at them knowing what was ahead. At first it didn't seem too bad, a slight incline, followed by a rest for drinks. No sign of Barry. I can handle this, I thought. 'Right, let's go,' said Barry, back from his bacon sandwich or whatever.

Somehow he managed to find an alternative way up, by car I think, but he would always be standing on the top of a hill above you shouting abuse: 'What's the matter with you lot,' variations on that sort of theme. I think he thought he had a squad of Chris Boningtons. You could barely stand up, let alone run up these hills. Mountain goats wouldn't have gone up there. Hang gliders were passing over the top of us and you could hear the pilots laughing. Where rabbits had dug their burrows, you were grateful for the footholds. All the while, I was passing prostrate bodies by the wayside. Kenny Lowe, who was a great runner but that day didn't have the power in his legs, had to be helped up.

You would get to the top of a hill where Baz had been sitting doing his impression of the Pillsbury dough boy only to find him gone. He was up on the next one telling us not to stop. By the time we finally did catch up with him, breathing out of our bums, we were too knackered to do him the damage we were threatening on the way up. Then, as his final flourish, he made us run back over the hills to the start, although at least it was mostly downhill. I was third or fourth back – not bad for me – and turned to see this great long line of people on the horizon. By then we had acquired reserves, youth team, schoolboys and anybody in the country who didn't have a contract. I think they had all come to Birmingham for a trial because they knew they had a chance with Barry as manager. There were hundreds of us. It was like a scene out of *Zulu* and gazing up in dread and wonder at all these figures you could just picture yourself at Rorke's Drift.

Insult was heaped upon injury when we got back to Dunstable Leisure Centre. We were changing, tucking into the bread rolls left out for us, then had them snatched out of our mouths because the staff had just received a fax from Birmingham saying that all refreshment had to be paid for by the players themselves. As with all people who seem to have a lot of money, Birmingham could be tight with it sometimes.

I thought it would be a relief, finally, to play football, but it wasn't: a programme of twelve friendlies was a season within a season. We played anybody and everybody – for the money, I suppose – including Bedford Town, Baz's old haunt as a player and the place where he still lived. In fact, we played them three times that season, whenever we had a free midweek, it seemed. I think the theory was that with so many players on the staff, Barry could field different teams for each pre-season friendly. Continuity and confidence does not work like that, though.

Rampant with a tie-up and in possession for Cambridge *(left)*, **before launching a long ball into the corner** *(Grus Darbo)*; **and** *(below)* **revenge against guru of the long ball, John Beck** *(Cambridge Evening News)*.

ed
ld
got
the
nier

was the man who
also killed off,
United's hopes of
the Premiership.
rm belief of Steve
e natural talent he
fied by the
anager's

me very good things"
We were a very
d highly intimidating
ainst.

ked on things
training to the point
cond nature to the

see the fear in the eyes
as when they came out
changing room at the
n. They knew we were
them around.

also a talented team,
e would have made the
e if only Beck had
a little and allowed the
press themselves in the
he field.

predictable, and the
the first division had
wouldn't have
'd been given a bit of
he system.
on about the system
ayers. But they were

REVENGE . . . Steve Claridge douses John Beck after United clinched the Division Three championship in 1991.

'You could see the fear in the eyes of a l

The beautiful game at Luton *(above, Josh Levy)* **and** *(left)* **Barry Fry in the dug-out at St Andrew's** *(Allsport).*

(Above) **The world's worst trainer** (Robert Hallam)**, but** (below) **always a favourite with the fans** (News Team)**.**

(Left) **Leaping for a ball against Oxford, at St Andrew's** (News Team).

(Below) **'Keep your head up, son.' Leicester v Stoke in the play-offs** (Empics).

Wembley wizards. *(Above)* **After scoring the winner in the play-off final against Crystal Palace** *(Jack Dawes)*; *(below)* **with Garry Parker, woolly bobble hat now replaced with an inverted baseball cap** *(Empics).*

The goal *(above, Empics)***, the celebration and the commiseration** *(right, both Sportsphoto)***, as Claridge clinches the Coca-Cola cup for Leicester . . .**

. . . **and the ten-foot-high effigy** *(Colorsport).*

When we did play anyone decent we were well outclassed. Liverpool and Manchester United had us chasing shadows but they were two weeks into training and all we had done was running; good practice for those matches, come to think of it.

But I was really looking forward to the season. The club was attracting big publicity with Barry, Karren and David Sullivan, and we were the biggest club in the division, the ground really beginning to look like a Premier League club's. Everyone was watching us and the whole place was buzzing with expectation. The trouble was, we were expected to mullah everyone in the division because we were the biggest club in it. Sometimes it felt like being a league club against a non-league one in the FA Cup, and we all know what can happen in those games.

Before our first match, at Leyton Orient, for example, one Birmingham supporter came up to me in the car park, showed me his betting slip and asked me to make sure the lads kept our win down to 3–0. It was never going to work out like that, though, as that match showed. Although I scored, we lost 2–1. Leyton Orient would later finish bottom of the league. All hell broke loose. The supporters were furious. One of the directors said publicly that he was concerned, and he now doubted our ability to go up. It didn't help when we then lost 2–1 at Shrewsbury in the Coca-Cola Cup in midweek.

We just sat tight, though, and a couple of wins, over Chester and Swansea – in the latter I got both goals in a 2–0 victory – kick-started our season. We also beat Shrewsbury in the second leg 2–0 to earn a tie with Blackburn Rovers. Still Barry insisted on changing the team and his favourite trick would be to throw on all three substitutes at the same time – half-time occasionally. By the end of the fourth league game, which we lost 1–0 at Wycombe, Barry had used twenty players. They were coming and going all the time. Someone once said that if you gave Barry five million quid he would go out and buy fifty players, so many did he know and remember from all his scouting missions, rather than five quality ones. That season at Birmingham I saw what they meant.

Barry was under a lot of pressure at that time. Soon after we lost to Blackburn in the Coca-Cola by a respectable 3–1 on aggregate, David Sullivan announced to the world that if Birmingham were not in the top three in November, he would have to think about changing the manager. Then, as if remembering the end of the last season, when we had some success with a settled side, Barry resisted the urge to tinker for a while. The result was a twenty-game unbeaten run over four months that took us to the top of the table and climaxed with a 7–1 win over Blackpool, in which I scored twice to take my tally for mid-season to 13. I remember the game well. Actually, we went a goal down and I missed at least four one-on-ones with the goalkeeper. To hell with this, I thought, and hit it from about 25 yards, though I am not the best striker of a ball in the world. To ironic applause from the fans, it just sailed into the net and put us 3–1 up at half-time.

We were now a really solid side. Ian Bennett was a terrific goalkeeper, Gary Poole had come from Southend to make the right-back position his own, Liam Daish was a tower of strength in defence, Mark Ward strong in midfield, and up front I was getting the goals. Ricky Otto also came from Southend in December to play alongside me, the gaffer insisting to a sceptical Karren and David that he was worth £800,000 as he would play as a striker and score pots of goals. He began with two in his first two games, against Cambridge and Cardiff, but then dried up. 'You're right,' Barry told Karren and David. 'He's not a striker. I'm going to have to play him wide on the left.' Don't ask me why, because he was different class when he arrived with his sweet left foot, but Ricky never really hit it off with the fans, despite scoring some crucial goals that season. Perhaps the big-club atmosphere got to him. Anyway, Barry soon asked for another £800,000 to sign Kevin Francis from Stockport County, and they reluctantly agreed.

Kevin, all 6 feet 7 inches of him (but a real gentle giant), was not everyone's cup of tea, yet I think that season he did more than anyone to get us up. Teams were beginning to come to St Andrews and put eleven men behind the ball, so we needed another option, and he duly provided it. It didn't matter how many men the opposition had in the penalty area, you could get it up to Kevin and he would knock it down. He also scored some important goals.

Not that it started that way. His second game was against Crewe and we lost 2–1. Afterwards Barry came into the dressing room and said: 'Kev. I have got to hold my hands up. I made a mistake. I've paid eight hundred grand for you and you're not worth eight pence. You're effin useless.' Kevin just sat there stunned, unable to believe what he was hearing. The rest of us just sniggered. It was typical Baz. He did it to me, he did it to everyone, just to see how they reacted to it. Most just sat and took it because Baz could be fearsome and Kevin was not the sort of lad to have a go back – only three or four ever did. I used to let him get on with it and have a go back when I had taken enough. Which, that season, came at Bristol Rovers at the end of March. We had been doing well, won two and drawn the other of the previous three games, and I had scored two goals. During the pre-match kickabout, Baz came over to me and told me that he was dropping me. I looked tired, apparently. 'Well, I always look tired,' I told him, 'but I don't feel it.' 'I'm the manager, I say you look tired, and you're sub,' he came back. 'If you don't like it, come and see me in the morning.' He put Paul Williams, whom he had just signed on loan from Crystal Palace, up front with Kevin. After an hour, we were one down and Baz turned to me and put me on. I scored 15 minutes from time and we got a point. 'I'll see you tomorrow,' I said at the end.

The next day, he told me I could go on the list and I would be away in a week. The next match, I was picked, nothing was said again and I never was on the list. But that was Baz. He liked it if you had a go back, it showed your passion. Also he loved the football manager bit, falling out with players, patching it up, all the intrigue. He lived for this man's world. He enjoyed getting angry, then having it all forgotten. He used to come in the dressing room and in his fury kick the ball bag. Gary Poole once had the idea that we should fill it with bricks. We never quite had the courage, but I think once the plaster of Paris from his broken foot had been removed, Barry would have seen the joke.

The great thing about Barry was that he never bore a grudge, and because of that I never really had a problem with him. Despite his two heart attacks (none while I was at the club, though he certainly put it to the test) he was a pressure-situation manager, loved all the big games and atmospheres and was a good motivator. He had a good eye for a player and could pick up bargains in the basement of the game.

But he was not the world's greatest tactician. In fact a postage stamp would probably have been enough on which to write down all he knew about tactics. He just liked anything that was fast and resulted in goals. He never took training, leaving it to Eddie Stein and David Howells, nor worked on drills or ball skills. I remember him at the training ground only twice, once in the winter in his shiny suit and snakeskin shoes, once in the spring in shorts and sandals. Both times were after bad defeats and he didn't take the softly-softly approach. At one point we thought he might be trying out some revolutionary new system when he told four groups of five of us to go and stand in the four corners of the pitch. 'Right,' he shouted, 'now run round the pitch.'

It was difficult for Eddie and David trying to organize sessions. When I first went there our training ground was a school pitch that had broken glass on it. Then we moved to the Civil Service ground by the M42, opposite the National Exhibition Centre, which was pretty good. The only problem was that we used to have meals there and one of the serving girls confided to one of the lads – big mistake, telling anything in confidence to a footballer – that she was having to go into hospital for treatment on genital warts. We stuck to the sandwich bar after that.

The other problem for Eddie and David was that there were 40 or 50 players at the session. If you did a shooting drill, you would get one shot about every hour. Then they would prepare a team for a Saturday only for Baz to come in, after a rethink on his own on the Friday night, and change it all around. I have known players who were in the team on a Friday not make the squad on the Saturday. Baz was the only person who picked the team.

Sometimes I thought he put too much pressure on players. He liked to wind you up to keep you on your toes, not only with his verbal jousting but also with his transfer policy. If you weren't strong-willed and didn't have faith in your own ability, you would go under. I was under a lot of pressure that season because we had so many strikers at the club, and he was continually being linked with more. I don't think there was a week during the season when it wasn't suggested in one newspaper or another that either I was leaving or a new striker would be arriving.

It had started even before that match at Leyton Orient. The night before the game I was alone in my room in our hotel. (In fact, I was always alone in my room in team hotels. Nobody likes sharing a room with me, although I always got on well with Steve Butler at Cambridge, and for the last few years I have been on my own. It isn't just that I am untidy but I always have trouble sleeping – I need only about four hours a night, except when I lie in, which can go on until early afternoon – and watch TV until three or four in the morning. Also, I tend to shout and swear in my sleep. My first room-mate, Billy Rafferty at Bournemouth, almost had a heart attack once.) The phone rang. It was Liam Daish telling me that a host of strikers including Jan-Åage Fjortoft, then of Swindon, Gary Bull of Nottingham Forest, Paul Williams, before he came on loan, and Guy Whittingham of Aston Villa had been seen in the lobby talking to Barry. It sounded feasible – you wouldn't put anything past Barry. I went for the wind-up big time and really hit the roof. In fact, only Gary Bull was there, though even that was unsettling on the eve of a match. After that I stopped worrying. Mostly Barry was keeping faith with me – I missed only two games that season – and instead kept changing my partners, of whom I had 23 during my time with the club.

The real pressure was getting into the team. Baz's policy worked in one way, in that it kept every player interested. You could be out as soon as you were in. Then again, you could be in as soon as you were out. Everybody knew they had a chance of playing. It didn't matter if you were the best-paid player or the worst, you always had a chance with him. Only a few didn't and he would get rid of them. Personally, I always found it a bit unsettling, and I thrive on being settled. I can't sit around, not playing; I vowed never again after Cambridge. I'd rather be in Aldershot's first team than Birmingham's reserves.

When you were in, though, Barry generally let you play with

freedom. He liked quick, passing football mixed with some direct, nice sweeping moves that he had been used to as an impressionable young player at Manchester United under Matt Busby in the fifties and sixties. By his own admission, Baz wasted his early career. He could have been a Busby Babe, instead of just a Baby Face, if he had applied himself more.

He would still shout and holler at you, though. By now my nickname at the club, coined by Eddie Stein, was Clegg after the dry-witted character in *Last of the Summer Wine*. I think Eddie may have got it wrong, though, and meant Compo, the wellied one whose dress sense some of the lads thought mine resembled. Anyway, halfway through one game, just after I had made a couple of mistakes, I heard Barry shouting from the bench: 'Cleggy, you're drunk.' I tried to ignore it but he turned to Eddie Stein and said: 'He's drunk, Ed.' So Ed started up. 'Don't be daft,' I shouted back. 'I haven't touched a drop all week.' 'Well, you're playing like you have,' he screamed. The next week after another game, Barry said to me in the dressing room: 'Cleggy, you are worth £4 million. You're the best holder-up of a ball in the country.' That's Baz.

Others had to learn to endure it just the same. There was one time he threw a cup of tea over Dave Barnett. Dave just sat there, licked his lips and said: 'I've told you before, Baz. I take two sugars.' Another time Barry looked at Peter Shearer and called him a coward. Now Pete was one of the hardest men in football I have ever seen. He would never flinch from a tackle. A coward he was not. Pete took it in, but did not react. He went away and brooded for about fifteen minutes and came back like a dinosaur whose tail has been tweaked. He got hold of Baz and said: 'Don't you ever call me a coward again.' Baz acted all innocent. 'What are you talking about? I never called you a coward,' he said. That, too, was Baz. He didn't realize what he was saying half the time, or the effect he had on players. Even Pete began to doubt what he had heard.

I'd had an indication of what Barry was like years previously. I once went to see a Barnet game where he was having a running feud with his midfield player Robert Codner, who later came to Birmingham on loan and recalled a night which was one of the most bizarre of his career. At the end of the game Barry had run on the pitch in his shiny suit, moccasins with the gold bar on them – he always was a terrible dresser – and made for Robert. It was a rainy night and the pitch was very muddy. As Barry shouted something, Robert, who had had enough of it all by then, turned round and hit him so hard, and with perfect timing in mid-stride, that Baz fell and slid six feet down the Underhill slope. As Robert walked off the field, Barry got up slowly, attempted to brush the mud off his suit, and made his way to the dressing room with as much dignity as he could muster. Once back in the dressing room, Barry punched Robert back, gold rings and all befitting the non-league Ron Atkinson, before being dragged off Codner. They met again later in the bar, Robert subsequently told me. 'Want a pint then, Robert?' was the only thing Barry said.

I wish Karren Brady could have been as direct as Barry. I'm afraid I never really had much time for her as a football person. She thought a club could be run like a tabloid newspaper but, apart from an element of showbiz and a lot of gossip flying around the place, it is a different business altogether. She also came from a wealthy background and I don't think she had much rapport with Birmingham's predominantly working-class support. And her reputation for hiring and firing staff did not endear her within the club. Barry also used to come in complaining about how she was interfering in footballing matters and transfers about which she knew little. Memos started coming down telling us that we could not do or say this and that, and that we shouldn't be out at this or that nightclub.

And there was also the delicate situation of her relationship with our Canadian striker, Paul Peschisolido. For a while it was mostly rumour within the dressing room, some not even knowing about the gossip, and the players used to talk about her as any set would about the upstairs staff. But we were soon reading in the papers that she and Pesch were an item. This made the players' relationship with Pesch very difficult and nobody could really be sure that stories would not reach Karren's ear through Pesch. It was a difficult situation for him sitting and listening to things being said about his girlfriend, and in the end he had to go – transferred to Stoke, although he was a nice lad. But you couldn't blame anybody for feeling uncomfortable when he was in the dressing room.

Karren could also be really tight when it came to money, as my contract negotiations with her always revealed. That season we reached the Final, against Carlisle United, of the Auto Windscreens Shield after seven tough games, during which I scored four goals. We were going down to Wembley for the Friday and Saturday nights before the Sunday game, but Karren announced that the club would only pay for one night at our hotel, the Swallow at Waltham Abbey.

Luckily Barry found a wealthy friend who sponsored our second night, saving us £40 each. We were also only allowed to buy – buy, not receive – ten tickets each for our friends and relatives. The club said that because tickets were so scarce, with our amazing 55,000 allocation out of 76,000 revealing the depth of our support, players should pay. That added to the dissatisfaction that we were on a bonus of only £1000, with all that extra revenue coming in. On top of that we each had to pay £8 for our club ties. Things like that knocked you back, undermined your morale. When you are doing badly, you expect nothing, you just get your head down and get on with things. But when you are doing well you expect to be treated well with things like perks.

And we did do well, winning the trophy with a goal by Paul Tait – a true Blue, having grown up supporting the club – in sudden-death extra time (as he had also done in an earlier match against Swansea). But even that couldn't pass without controversy, this being Birmingham City. We only knew about it the next day when we read the papers but Paul had lifted the trophy wearing a T-shirt saying 'Shit on the Villa'. We didn't take any notice but everybody over-reacted and there were a lot of problems at the club for a few days. On one side of the City he was the dog's bollocks, on the other he was like Salman Rushdie. He had to keep his head down for a while, which was a bit difficult for Taity as he is a very social animal. Actually, he was beginning to settle down a bit and stay in a few nights by then, although this may have had something to do with the fact that he was barred from quite a few places.

Wembley was a fantastic experience, as had been our FA Cup tie against Liverpool earlier that season. During that game I really enjoyed my tussle with the excellent John Scales, one of the best defenders I have ever come up against, and we drew 0–0 at home before losing only on penalties at Anfield after a 1–1 draw. Promotion remained the goal, however. We had gone into the Auto Windscreens Final on the back of a good 3–1 win at Plymouth, in which I had scored twice, and three days after it we had a massive home game against Brentford, our rivals for the title and automatic promotion. We duly won 2–0 in a tight game in front of more than 25,000 people – the only blemish being a bad injury to Kevin Francis that kept him out for months – and it looked like we were almost home. Then came two deflating home draws, against Brighton and Bradford.

Yet again I was involved in a last match that had so much riding on it, this time against Huddersfield, who were a top-six team. It was a tense day and I didn't help our cause early on when I missed a simple chance, my shot hitting a defender on the line. But I made amends by turning home a close-range shot to set us on our way and we eventually came through 2–1. It was my 20th league goal of the season – 25th in all competitions – making me the first City player since Trevor Francis, twenty years earlier, to reach that figure in a season, and won me another supporters' award for Player of the Year.

I was particularly pleased because I played the end of the season with two broken ribs. About six minutes from the end of the game at Plymouth I was kicked under my chest and couldn't breathe. The bench were giving me stick but I stayed on and scored, although I knew something wasn't right. Every game after that I was in a lot of pain whenever I was elbowed or pushed, but I have always liked to play if I possibly could and I struggled on with a cricket thigh pad strapped round my rib cage.

It had been a great season. We were averaging crowds of 20,000 in the Second Division, with very few away fans, the stand was finished and the ground was looking magnificent. I really felt that I was at what would very soon be a Premiership club, that finally I had a chance of getting in with the big boys. I really felt we could make an impression in the First Division.

By now I had also managed to win the fans over. I could make mistakes and they wouldn't get on to me, unlike with others. Because I worked hard – though I always reckoned my game had plenty of thought and skill to it as well – and always gave my utmost, these passionate, working-class people recognized a kindred spirit. A player does know that if he gives 100 per cent, fans will sense it and give him the benefit of the doubt. They may have a pop now and then in the heat of the moment but generally they will be on your side. I was settled and happy and enjoying the training, travelling up every day from Luton in good company in a car with Jonathan Hunt, Gary Poole and Chris Whyte from London. I had my gripes but things were on the up and getting better.

I was still having problems getting what I thought was a decent pay rise, though. My money automatically went up £5000 that summer because we had been promoted but some players who had been on low salaries went up £50,000 and I was still on less than many although I was supposedly one of the top players at the club. Even players coming in on loan for a couple of months were getting bonuses of more than £5000 on top of their wages. It was hard to stomach after what I believed I had been promised. I asked for a rise of £20,000 and in the end we compromised on £10,000 but I was still a bit put out that eighteen months after joining the club I was only on £15,000 more than I had been at Cambridge.

Still, I was happy to stay, despite the pre-season prospect of Dunstable Downs again. The feeling was that we could go through the First Division and straight into the Premiership, confirmed when we comfortably beat two of the favourites, Ipswich and Norwich, who had both just been relegated, early on. We weren't playing great football but we were working hard for each other and things were going well. As ever, there were comings and goings, the departure that I noticed most that of little Jose Dominguez, our Portuguese left-winger, back to Sporting Lisbon. What a player he was.

When he first arrived we thought he must be another of Baz's non-league triallists. Then one day we were training on a muddy old school pitch. No one could stand up, no one could move the ball. Jose could not only stand up, he stood out. He was like Maradona, just dancing past people. We couldn't believe it. Which non-league club was he from? What was he on? He sat a couple of people on their arses and just looked really special. I have never played with anyone as skilful. Mind you, he was better to watch than to play with because he wasn't a great crosser. The other problem at that time was that he was not very strong and couldn't last ninety minutes. He did his best work as a substitute. Still, First Division defenders could be grateful that season that he had returned to Portugal.

In October, we went on a four-game winning streak and I hit a seam of goals, scoring both in a 2–0 win over Southend and another in a 1–0 win at Portsmouth. I also got two more in a 3–1 win over Grimsby and one in the 2–1 victory over Port Vale. We had a chance to go top of the table in early November when Millwall visited; they were top and we were second.

A full house of more than 23,000 was in place on a rousing day but unfortunately it degenerated into a fiasco of crowd trouble. The club put Millwall fans underneath ours at the Railway End, despite the reputations of both sets. Now some Birmingham supporters, even if they may not start the trouble, are not shrinking violets, and bombarded the Millwall fans with missiles. There was also a claim that one of the Millwall subs was punched by one of our fans, although it was never proven. The 2–2 result was almost forgotten. I do recall driving home that night and passing the Millwall team coach on the M1. There were only two windows left unstoned, at the front, and all the players were huddled in the first four rows. A different form of air conditioning, I suppose, though it was winter. It did mar what could have been a very good day for us.

We also had some much-publicized trouble on our sorties in the Anglo-Italian Cup. It started when we went to Perugia and two of their fans and two of ours were involved in a knife fight. We saw one who had been stabbed staggering into the ground when we were kept behind for forty-five minutes at the end of the game. Then came the game in Ancona, trouble this time involving the players.

It was a nightmare trip, with all sorts of delays and the club deciding we should go out and back in a day. I had had a touch of flu and wasn't playing but travelled anyway and watched the game from the bench. It started well enough but degenerated after we went 2–0 up. Their coach put on a lad who just started kicking and elbowing everyone and then our lads started to get a bit naughty as well. At one stage their coach ran on to the pitch and tried to throttle Paul Tait. He was hauled off but not before some argy-bargy began between the two benches. We won the game, holding on after they had pulled a goal back, but nobody cared, trouble was brewing and it was a case of looking after yourself.

Their coach was now having a go at anyone he could find and violence had flared in the stand. I was due to go for a run round the pitch with the physio while the teams were getting changed but that was all but forgotten, to my relief. Our lads decided to walk down the tunnel together, led by our coach David Howells, who is the sort of, shall we say, upright, formidable figure you wouldn't want to mess with. I was watching the fighting in the stands when all of a sudden I heard a big noise in the tunnel. Trapped, with the chaos behind me, I decided my place was in the tunnel but luckily I arrived after the first blows had been struck.

It seemed the Italians fancied a go but bottled it when they were confronted, confirming my opinion, formed as an 18-year-old that time when I went to the European Cup Final in Rome to follow Liverpool, that our footballing cultures just do not mix; we have such different temperaments. They spit and pull your hair, which English players do not do. Our thuggery is more open and honest. Anyway, it was pandemonium when I got there with people being pulled apart. Their coach was now on the ground. At that point we didn't realize the extent of his injuries, which the papers said were life-threatening, but at the end of the day our view was that he had brought it on himself. I remember one of the lads had taken my boots into the dressing room for me. When I got them back, there was blood all over the studs.

The talk was that Liam, David Howells and our defender Michael Johnson, who had all been earmarked as being at the front of the trouble, might face prosecution in Italy, though it eventually quietened down. Johnno was especially worried and the lads did nothing to ease his fears. He fancied himself as a bit of a looker did Johnno and was subjected to comments along the lines of: 'Ooh, Johnno. All those pretty boys in the Italian jails are going to love you. You're going to be their little plaything.'

Our experiences in Italy seemed to sidetrack us from the league and we began to slip. I also missed six games having had a kick in the back which put it out. I had to go to Sheffield twice a week for six weeks to have it manipulated until it was back in position. We did, though, have a great run in the Coca-Cola Cup, reaching the semi-finals after a mountain of tiring games, including Tranmere in a replay and two tough legs against Middlesbrough.

We had beaten Plymouth and Grimsby in the first two rounds before encountering Tranmere at home, having to hold on to a 1–1 draw after Gary Cooper had been sent off. Then up north we played well for a 3–1 win, Ken Charlery coming on and scoring twice. Ken was another of the many strikers at the club, a good player but one who could not always handle Barry's hot and cold moods. We were a bit fortunate to get Middlesbrough in their bad patch after a good start. At the Riverside Stadium, Juninho, who had just signed, was in a different class, but we held on for a 0–0 draw, with me playing in a five-man midfield for a change. At home, we beat them well 2–0 with the Brazilian uninspired as Boro ran out of ideas.

Then, after we had beaten Norwich 2–1 at home in a fifth-round replay, I was staggered to find out that Birmingham were trying to sell me before the first leg of the semi-final against Leeds United. At the start of the season, when I was asking for more money, Karren had written me a letter saying that just because I'd had a few good games during the promotion season didn't mean I could hold the club to ransom. That, I thought, just showed what they thought of me after what I had done for them, and this new episode only reinforced it. After the contract negotiations, I wasn't going to ask for more money again, but just sit tight till the end of the season and take it from there. They knew I wanted to stay and wanted an extension to my contract, which would then have only a year to run.

We were playing Sheffield Wednesday in a friendly behind closed doors in early February – one of the manager's ideas; he was trying to sell players and the Wednesday manager David Pleat was interested in a few of us – when Barry pulled me to one side. He said that they'd had a lot of bids for me but that he wanted to keep me and put me on the same money as Mark Ward, who he was letting go on a free transfer. I think Barry might have been a bit worried that Mark might be in line for his job. If he couldn't get the pay rise for me, he said, he would let me go and I would get it somewhere else. I thought that I couldn't lose here. This was a promising turn of events.

The next thing I knew, the papers and Clubcall telephone service were full of stories saying that I had asked for £3500 a week and would leave the club if I didn't get it. This despite having had two pay rises in eighteen months. Karren said that the stories didn't come from her. So too did Barry, but he was on Clubcall saying that I had played my last game for the club. I was horrified. This was not the real picture – I felt I was being cast as the villain of the piece. One Friday night I was listening to a sports show on the local radio in Birmingham and the presenter was saying I should stay for the semi-final at least and that the situation was of my making, an example of money gone mad. I felt so incensed that I rang in to say they had got the story wrong and one day my side of it would be heard properly.

To this day I don't know why they wanted to sell me but it was true that Barry could love you one minute and want to offload you the next. And I was in a bit of a barren spell, due to an illness that was beginning to grip me. Perhaps they thought they could cash in before the transfer deadline in March. Perhaps they wanted Paul Peschisolido, who was by now married to Karren, back at the club in my place. Anyway, there was a lot of talk that Leicester City's manager Martin O'Neill, who had tried to sign me in the past, had been in for me and Barry asked me to go and talk to him. It looked as if a deal had already been done and I was by now very disillusioned. It was all getting on top of me. I hadn't scored for a couple of months and now, before their biggest game in twenty years, it seemed I wasn't wanted. On top of that I was feeling debilitated by a virus of some sort, which I thought at first was flu but became much worse. I knew something was wrong when my stamina, one of my assets, was failing me. Players I always beat would be passing me on long runs. Later it would be diagnosed as a thyroid illness.

When I came back from talking to Martin O'Neill and said that we weren't far apart on the terms, Baz asked me what I was doing getting into such detail because no fee had been agreed. Why was he letting me go and see him, then? Then he dropped me for a match at Stoke, saying that a transfer was imminent and he couldn't risk me. That Saturday night I also spoke to Martin.

Soon it got more ridiculous at the club. The following Monday at training, Eddie Stein asked me if I wanted to take part in a first team v reserves match. With a transfer pending, I declined, not wishing to get injured, so I was made to run the line and retrieve the ball when it went out of play. I could have refused, I suppose, but by now I was keen not to upset the applecart. I had conditioned myself that I was leaving when Barry suddenly put me in for the next game, away to Barnsley on a bone-hard frosty pitch that midweek. I had a nightmare. I was not in the right frame of mind.

By now I was totally shattered with the medical problem having taken a hold. I had no energy and just wanted to disappear. Little chance of that, though. I was trying to keep a dignified silence but supporters were coming up to me in the street calling me a greedy bastard, calming down only when I told them what the real situation was. Karren rang me up and said that if I didn't leave the club Barry would resign and I would rot in the reserves. In ten days I had become the club's worst enemy, it seemed.

But Barry put me in against Leeds, even though I was ill and the transfer to Leicester was pending, and I had a really good game, even though we lost 2–1. In fact, I was ITV's man of the match. I was just playing on adrenalin, all hyper because it was such a big fixture. I still had a lot to give the club, I felt, confirmed when I went home that night and watched a video of the game, in which we had taken the lead through Kevin Francis but committed two silly defensive errors.

Then I was asked to go and see Martin O'Neill again the following midweek, the day we were playing at Crystal Palace. My Leeds performance had meant nothing and I was still *persona non grata*. I felt a little bit under pressure at Leicester, with Martin wanting my signature, but I was still not quite ready to sign. I was being offered £3000 a week and wanted £3500. Martin said he would phone me at my Luton home at 4.30 that afternoon but when the call didn't arrive by 4.45, I set out for Selhurst Park.

The traffic was so bad around those streets of south London that I started to get worried. I rang Barry on his mobile to tell him I was nearing the ground and would be there in ten minutes. He said not to worry. Then when I got there, he told me I was too late and that he had taken my name off the team sheet, though I don't think he intended to play me anyway. He asked me how the day went. Only later did I find out that he went from there to do a radio interview in the press box saying that I had turned up late and branded me a disgrace. The next day I picked up the *Sun*, which quoted Barry as saying that I had refused to play and that he would be fining me. I was furious and rang the *Sun* reporter to put my side of the story for the first time in two months.

I was sure that I would now be off within days – after all Liam Daish's move to Coventry at the same time took only a few days – but every time a deal appeared to be sealed, there would be a hitch, both sides claiming the other was holding up the agreement. I was piggy in the middle. Barry, though, had one more game for me, the second leg against Leeds, and I was glad to play in such a big match, even though I was feeling listless with the illness. I felt I owed Birmingham's fans one last tilt at the windmill.

Even after the first-leg defeat, when we didn't do ourselves justice, we felt Leeds were beatable. They really were no great shakes. And as Aston Villa were to be the opponents in the Final we really wanted to do it. I might have known what was in store for me early on, though, when I collided with the referee. Somebody knocked a ball and I ran after it and he just appeared from nowhere, leaving us both in need of treatment. Then I missed a penalty that might have put us back in the tie, which just about summed up my final period with the club. Even then the fans were still good to me, chanting my name and never jeering me.

We lost 3–0, 5–1 on aggregate, and it was an anti-climactic end after everything I had been through with Birmingham. In many ways I still did not want to leave. This was a big club and the grass is not always greener elsewhere. I had got fed up with moving. After my relationship with them, I couldn't understand the rush to get me out – though things soon became clearer when they re-signed Paul Peschisolido, having denied for so long that that was the reason. Yet in the summer, a few months later,

Pesch was gone again, to West Bromwich Albion. Birmingham City move in mysterious ways.

Not so mysterious, perhaps, was their decision at the end of the season to dismiss Barry, who promptly bought into Peterborough as director of football so that he could continue his life's work of buying and selling players. He had been perfect for Birmingham for a long while, bringing them promotion and publicity, but it seemed he had gone as far as he could with the club. Now I think David Sullivan, having become well known for more than soft porn, wanted to take the club upmarket.

The appointment of Trevor Francis, the club's favourite son, was a marvellous one, just what the club needed. He immediately went out and got in some quality players in Steve Bruce, Barry Horne, Gary Ablett, Mike Newell and Paul Furlong, instead of buying in quantity. It was funny, but David Sullivan, not the ducker and diver or bully he is sometimes portrayed as but actually quite approachable, once asked me who he thought we should get to play alongside me, and after mentioning Alan Shearer, I gave him the names of Newell and Furlong.

It was funny, too, how things were to work out for me. Birmingham still owed me the final instalment of my signing-on fee for them and, on advice from Brendan Batson of the PFA, who I was now using rather than an agent, I insisted on getting it before I would leave. Luckily they agreed, as I sensed that Martin O'Neill was not a man to be messed with for much longer, and finally I agreed to his pay offer and the move to Leicester. My money would go up another £500 a week if they made the Premiership but I thought that was a fat chance. My best chance had gone with leaving Birmingham, I thought. Another club, another chapter, though. And another twist to my ever-unpredictable career.

8 The Million-Pound Goal

Steve's transfer to Leicester City finally went through at the beginning of March, 1996. Geographically it was a sideways move, from West Midlands to East, and metaphorically it seemed the same for a player forever on the fringe of the big time. Like Birmingham, Leicester were another of football's middle and middling men, a club with a great future behind them. They'd had some stirring days in another era when the game was less about the size of television deals and the chairman's financial portfolio, reaching three FA Cup Finals in the sixties – all lost – and winning the League Cup in 1964. They had even been up into the Premiership. Naturally they came straight back down after one season. Now they were little better off than Birmingham, hovering in that First Division zone between mid-table and the play-offs.

Steve's start, for a club looking for a second wind to the

season, was a typical one for him: fraught with problems as he settled slowly. He made his debut at Ipswich and found himself in a team 3−0 down by half-time. They eventually lost 4−2. It was the first of six games that saw him failing to score, four of which Leicester lost. Looking up from ninth position, promotion was a distant dream. Twice Steve went in to see the manager Martin O'Neill to apologize for his form.

It was, he explained, due to the continuing illness that had beset him during his latter days with Birmingham. He was forever tired, sleeping half the day and still not feeling refreshed. He lost sensation in his feet. At first he dismissed it as mere flu, then the stress of the transfer and the wrangles with Birmingham. The idea of ME or even multiple sclerosis darted alarmingly through his mind. It was a worrying time. Leicester sent him for tests on a trapped nerve. Finally it was diagnosed that the tablets he had taken for his heart defect for the previous eighteen years had caused a serious side effect. New medication was prescribed.

O'Neill was considering dropping Claridge for a make-or-break game against Charlton Athletic at the beginning of April. 'I know what you are going to say,' he said to the manager as he came to him a couple of hours before the night's kick-off. 'But don't. Just give me one more chance.' O'Neill agreed. Then, in the first half, Steve received the ball some 20 yards out and looked up. He saw Charlton's nervous Australian goalkeeper Andy Petterson, who was playing his first game of the season, slightly off his line and curled the ball unerringly and expertly into his left corner. 'The relief on his face was plain to see,' his partner up front, Iwan Roberts, remembers. 'It was a cracker of a goal and all his frustrations came out in his celebration. He was running around the place like a madman.'

Leicester clung on to the precious goal for a 1−0 win and the kick-start to the promotion push they needed had finally arrived. They won their last four games with Steve scoring four goals, including two against Oldham Athletic and another against Birmingham City on an acrimonious day for him at Filbert Street, when an appreciative gesture to his own bench was misinterpreted by Blues fans who had previously idolized him. It maintained his remarkable record of scoring against his former clubs. Leicester finished 5th and squeaked through their play-off against Stoke City 1–0 on aggregate to set up that glorious – both for Claridge and the club – May day at Wembley against Crystal Palace. Finally, after a professional career spanning fourteen years, his epic shot had earned him one at the Premiership.

'He deserved it,' says Iwan Roberts. 'He is a terrific player at holding the ball up and certainly the hardest working I have ever played with. Everybody took to him at Leicester. He was great in the dressing room, a very bubbly character. He took a lot of stick for the way he was but always took it in good heart.' The Claridge characteristics remained in evidence within Filbert Street, it seems. 'If you started training at 10.30, Steve would arrive in the car park in a mad dash at 10.29 a.m. having driven up from Luton with his tracksuit and boots on.'

The final ironic statement on Steve's timekeeping had come, in fact, just before his move to Leicester. 'My watch is tops,' he was quoted as saying in a Birmingham City match-programme advertisement for club timepieces. 'With its classy design and easy-to-read watch face, I am never late for training or Blues games.' It was published four days before his late arrival at Crystal Palace.

During pre-season for the 1996–97 season, Roberts, who had by now moved to Wolverhampton Wanderers and took his family over to the Black Country to stay in a hotel with him, lent Claridge his house in Leicester for a week. 'When we got back the place was a tip. He hadn't made the beds and all the pots and pans were still in the sink. Steve's explanation was that he had to be at Filbert Street at 6.30 a.m. for a trip up to Scotland for a pre-season friendly. Actually, I was told by one of the other players that Steve must have interpreted that as having to get up at 6.30 because he was late at the ground. We didn't mind, really. My wife thinks the sun shines out of Steve's backside.

'We also asked Steve to feed our dog and two cats. He said he had fed the dog but not the cats. Apparently about fifteen of them on the estate would turn up and he didn't know which two were ours. I later kidded him that one of them had died from starvation and the lads at Leicester were winding him up about it as well. He fell for it for a while.'

'Did I hell,' says Steve now, with a smirk.

'We knew about his gambling but to my knowledge in the time we were at the club together he didn't go in the bookie's,' says Roberts. 'I don't think he's into it as much as he used to be. Mind you, when lads at the club said they were going to the bookie's, he sometimes would take on the bet himself. A little bit about his gambling appeared in a Sunday paper on the day of the play-off final and the lads were getting stuck into him, light-heartedly. "How could you do it on this of all days," we were saying. "You'd better make up for it on the pitch." He certainly did, didn't he?'

Indeed, Steve will forever go down in Filbert Street folklore for it, along with the goal that won the Coca-Cola Cup against Middlesbrough the following season. Then again, he has clearly imprinted himself, in one way or another, on all the clubs he has been with, from Portsmouth, through the non-league, via Cambridge and Birmingham to the Premiership.

In speaking to the team-mates and managers who have come into contact with him, their response (to a man) to his very name brought a chuckle. All then agreed that having served his time and paid his dues at every level of the English game, he deserved his days in the sun at the age of 30. Some, it has to be said, doubted whether he was of that quality, notably David Pleat. Perhaps he was the archetypal player for that Division One and a Half. Was there, as Steve insisted, more to his game than honesty and hard work? Would there be evidence against the great names of the English game of the thought, skill, creativity, ingenuity and sleight of foot he insists he also possesses? All would be intrigued to find out. And despite all the comic moments of his career he does deserve to be taken seriously as a player.

As might be expected, it began inauspiciously. After putting in a week at the club's pre-season training camp at a marine base in Devon, where this time he let a proper, if severe, barber shave his head, he missed most of Leicester's friendlies and their opening game at Sunderland with an ankle injury. He finally made it in the next midweek, on the winning side, too, as the Foxes beat Southampton 2–1 at home with two goals by Emile Heskey. 'I didn't have a great game but it was a great night,' he said. 'I had made it at last.'

It was far from over yet though.

Leicester had paid £1.2 million for me – not bad for a lad from Portsmouth who was told he was not going to make it – but I soon turned into a million-pound flop. They thought they were buying goals but I simply couldn't buy a goal when I got there. This on top of thirteen games without one for Birmingham. It always seems to happen to me; I am just a bad starter. This time, though, there was more to it.

We were 8th in the First Division table when I joined, and by half-time of my first match we were 3–0 down to Ipswich – me playing with my shorts on the wrong way round. Eventually we lost 4–2. It was live on the television and I can only imagine what people were thinking about me. And though we won our next game, 2–1 over Grimsby, the play-offs looked a long way away. In fact we slumped as low as 9th after a 2–0 defeat by Sheffield United. That was my sixth game for the club and I still hadn't opened my account. It was now nineteen games and counting since my last goal.

I had been feeling tired all the time and just didn't understand why. I knew there was something wrong with me but I thought it was just a hangover from the Birmingham experience. That had brought the illness out, it was to transpire. Just standing up for training was a real effort (no change there, then, some of the players who have known me will tell you). I was cold all the time and always seemed to be moving in slow motion. I know I may not be the brightest lad in the world but I was even more backward in my thinking, and my reactions were sluggish.

After one game, a 2–0 home defeat by Ipswich again, I asked to see Martin O'Neill and told him that the club was not seeing the real me and I couldn't understand why. He was sympathetic but I think he was beginning to have doubts. At half-time of another game against Millwall he told me that my first three touches were a disgrace. I replied that I might be in a position to have agreed with him if I could have felt them. I had pins and needles in both my feet. In that game I made one early run and just felt finished after it. Against Oldham I shouted to our physio Alan Smith that I wanted to come off. That certainly was not like me; I love playing. I also needed up to twelve hours' sleep a night, where often I had been used to getting by on four, and still woke up feeling I could sleep another twelve.

The medical I had before signing showed up none of this; those just deal with the soundness of your limbs. The club thought I had a trapped nerve but decided after the Sheffield United game to send me for a blood test anyway. It revealed that I had a deficiency of a vital chemical called thyroxin and consequently my body was packing up on me. Apparently, it mostly affects 55-year-old women, so I either had the body of a 55-year-old woman, which some fans might have thought was the case, or something else had brought it on. It turned out to be that the tablet I had been taking for my heart had knackered my thyroid gland after fourteen years. Some of the papers said I had ME – the yuppie flu. When I told the doctor how little money I had to my name after all the gambling he agreed that, no, I couldn't possibly be a yuppie.

It was a huge relief to find out finally what was wrong with me. I was beginning to worry and I am only thankful that I read an article about the magician Wayne Dobson and multiple sclerosis the day after the diagnosis. I seemed to have all the symptoms he was talking about.

The medication I was prescribed helped almost immediately and I went to the gaffer to ask him not to drop me for our next match at Charlton, something I knew he was considering. With Charlton in the play-off picture too, it was going to be a really big game, one we needed to win if we were to make it ourselves.

For the first 20 minutes I was still diabolical. Then, all of a sudden, the ball came to me and from about 20 yards out I curled a shot into the goalkeeper's left corner. All the frustration of three months without a goal came out in my celebration afterwards. It proved enough to win us the game and I and the team were on our way. I remember being asked by newspaper reporters afterwards if I thought we could make the play-offs now. 'Well, I've got more chance than I had with Aldershot,' I told them.

We were back in London the following Saturday and recorded another excellent 1–0 win over Crystal Palace, who were looking good for promotion themselves. It moved us up to 7th in the table. Another wobble was to come, with a 2–1 defeat by West Bromwich Albion and only a point from a 1–1 draw with Tranmere, but there was certainly more confidence in the team. Mine was returning after what had seemed a lifetime. Suddenly the crowd, on whose support and energy I have always thrived, were on my side.

We just had to beat Oldham, who were themselves desperate for the points, at home after the Tranmere draw. You could sense the tension in the crowd and their early frustration as Oldham played five at the back. In these pressure situations you find out the real character of a team. At Luton we had folded when the going got tough; Leicester was the opposite – we found a way through. I got a tap-in which seemed to get us going and also grabbed the second in a 2–0 win with one of my better goals, having held off two defenders trying to foul me on a long solo run before slotting it past the goalkeeper. Three days later in another home game against Huddersfield I got the second in our 2–1 win and we were now 6th.

The following Saturday we had our third home match in a row. Against Birmingham City. Almost 20,000 were in the ground and I felt the hugeness of the occasion more than anyone, I think. As I said hello to some of my old team-mates in the corridor before the game, all I got as response was silence. They were trying to wind me up but it wasn't going to work. So too were the Birmingham fans when I ran out. I never realized they were such a rich bunch. They were waving fivers and tenners at me as their way of telling me that I had gone for the money, which, of course, had never been the case.

The Italians have a saying in football about the 'immutable law of the ex', which means that a player will always score against his former team, as if by destiny. It had always happened to me previously and it was to do so again that day. The game was goalless when Scott Taylor volleyed a ball across their area, Emile Heskey flicked it on and I met it at the far post and headed it back across goal and into the net. I was delirious and pointed to Martin O'Neill on the bench, mouthing my thanks that he had kept faith with me. It almost caused a riot as the Birmingham fans interpreted it as a derogatory gesture. It was certainly not one directed to such a great set of fans, though I felt like making one at the club's management.

We went on to win 3–0 and confirmed our place in the play-offs, in 5th place in the table, with our fourth consecutive win, 1–0 at Watford – comfortable, despite the scoreline – on the last day of the season. We were pitted against Stoke City with the first leg at home but could only draw 0–0 in a disappointing game. After all the struggle of the last two months, it looked as if our chance had been blown in ninety minutes, which can be the nature of the play-offs. But we went out and won the return 1–0 at the Victoria Ground, with a goal by Garry Parker to take us to Wembley for a final against the favourites Crystal Palace, who seemed likely to win automatic promotion for much of the season.

What a day, what a match, what a moment it produced for me. Though I have got used to play-offs after a career spent in the lower divisions and see them as an occupational hazard which adds interest for the fans, people in the game still moan about the concept, claiming that a knock-out competition should not be used to settle a league; that play-offs are essentially a lottery. In England's most famous stadium on a glorious spring day, though, I felt like all my six numbers had come up.

Against the run of play, Palace took a first-half lead through Andy Roberts, but we always knew that if we kept up the standard we were showing we would get an equalizer. That is the theory, but as time ebbs away you do begin to doubt it. Then, 14 minutes from the end, Garry Parker, our man of the match with his running and classy passing, grabbed the goal. Relief coursed through my body. We felt the stronger as the game went into extra time but a winner just would not come. It looked like we would have to do it via penalties.

Then, with no more than a minute to go, we won a free-kick on the right. I was yelling at Garry not to take it. I was so tired that I was having trouble getting up into the penalty box. My legs were cramping. So when the ball came out half-cleared, I wouldn't normally have been there, some 20 yards out, but further forward up in the penalty box. Later I think I said something daft to Gary Newbon on television about having shinned it (I wanted to get the interview over with and join in the celebrations) but actually I caught the ball perfectly, hit it as sweet as a nut. It just sat up and winked at me and I hit it. I knew instantly it was going to sail into Nigel Martyn's left corner. He was unlucky: as it was on its way, I noticed a Palace defender run across his vision. Frankly, though, I didn't give a damn. I didn't see it hit the net. I didn't need to. I was just off and running in celebration and saw our fans go up as one. Suddenly I had all sorts of energy – how far away now seemed the thyroid problem – and, if Muzzy Izzet hadn't caught me, I would have been out of the gates and round the Wembley concourse.

I can still picture the goal in my mind and hardly need to replay it on the video, as I have done a thousand times since. It still gives me goosebumps. Whenever I am feeling a bit low about my game, I just put it in the video to show me what is possible. I have heard people say after they have done something big in their lives that they can now die happy. I now know that feeling. No one can ever take it away from me and after all I'd been through that season, all I have been through in my career, it was the best possible crowning moment. It was my fifth goal of the promotion drive and illustrated again a curious aspect to my career; how my goals seem to come, like London buses, in bunches. You wait a couple of months and five come along all at once. I don't know why this should be. It doesn't seem to be anything to do with confidence as, by nature, I am always an optimistic player. But it does seem to reflect the way I live. When something has gone wrong, something good always comes along. When things are brilliant, something comes along to knock you down. It is as if I need the extremes. I don't seem to be happy amid the average, the mundane.

Leicester deserved their success too. There had been some good passing football, through the likes of Muzzy and Garry Parker, and some incisive direct play; a mixture which suited me well. Mostly, the run-in was to do with character, though.

And Martin O'Neill deserved promotion as much as anyone. He had faced a difficult task in replacing Mark McGhee and finally got it right after his unhappy experiences at Norwich, having rejected the chance to manage Nottingham Forest when he was at Wycombe Wanderers. Leicester's long-suffering fans, having seen the club lose a play-off final to Blackburn then witnessed promotion to the Premiership before instant relegation, had also endured Brian Little's departure to Aston Villa and were in need of a boost.

For me, after a decade in the lower divisions, the Premiership represented the grail and at 30 I had found it, though I did have a false start. After pre-season training at a marine base at Torpoint in Devon – a doddle for me after Dunstable Downs – I sustained an ankle ligament injury and missed the first match of the season at Sunderland. I did make it the following midweek in our 2–1 win against Southampton. Then, in our fourth match, came my first Premiership goal – one that I can number among the best of my career. Intercepting a pass from Des Walker I ran on a few strides and cracked a left-foot drive – which Sky TV timed at 51.5 miles per hour, hit from 26.2 yards – over Kevin Pressman into the top corner of the Sheffield Wednesday goal. Unfortunately we lost 2–1, but I had now scored in every division of the English game.

There could hardly have ever been a better time to be a top-flight footballer in this country and I look at my young team-mates like Emile Heskey with envy. If they apply themselves the sky really is the limit. With all the attention the game gets, on Sky television, on radio, in the papers and all the glossy magazines about now, everything is just so exciting. Footballers used to look up to pop stars. Now it seems all the pop stars want to know footballers. Be them, in fact.

I enjoy the rest during the summer, but I miss football badly. There doesn't seem to be anything worth reading in the papers except who's going where in the football world. On TV, I might watch the second week of Wimbledon or the Open golf but otherwise I can't wait for the football to come back on. Two weeks without it is enough for me. In fact, I spent my holiday in Greece watching Euro '96. In the winter, I might watch some snooker or darts, or tune in to the Five Nations rugby, but no other sport contains the same passion.

For me it is the be-all and end-all of life, which may be why I get on so well with the fans, many of whom have the same attitude. Suddenly I was enjoying lots of attention, after going unrecognized for so long at Aldershot. I signed more autographs in one afternoon than I did my whole time there. One bloke came up to me and said, 'All right, super Steve?' It seemed my name had changed. It gave you a great buzz to hear Filbert Street chanting: 'Super, super Steve, Super Stevie Claridge.'

There were downsides to being a premiership player, of course.

With all the interest in players, anyone and everyone can get turned over in the tabloids these days. And when you are out, people do stare at you. You wonder if they are listening to what you say, watching what you do and what you are drinking, ready to ring up the papers. At Birmingham it got out of hand with people criticizing me for being greedy, but once I had explained my position to them they usually calmed down. I never had it as bad as Ricky Otto, who really used to get some stick. Others want to tell you what's wrong with the club, or expect you to tell them what's going on inside it. Often I don't agree with them and can't give them any confidences but I try to accept that they have their views and listen to them.

With all the money the fans pour into the game these days, it is the least they deserve, apart from 100 per cent effort on the field, that is. Sky were pouring in £185 million a season into the Premiership – £9 million per club – and there was talk that the next TV deal could be more than £1 billion, but they would not be able to do so if it weren't for people paying subscriptions. Then there are the increasing attendances, not to mention all the merchandise the fans buy.

It enables us players to earn salaries that stars of yesteryear can only shake their heads at. Roy Keane's latest deal is worth £50,000 per week but, though many people mutter about such sums, I don't begrudge any player whatever deal he can negotiate. You can be sold without warning, you can break your leg, and you are lucky if your career lasts until you are 35. So you owe it to yourself and your family to get what you can while you can. It doesn't mean I am going to feel any less excited by the game itself or give less to it. It's just being a professional.

I wouldn't worry if a big foreign player at my club was on a lot more than me. I would say good luck to him, though clearly it creates friction if one player is on £25,000 and another on £2000. I do worry, however, if a player who is worse than me is getting more. It is up to you to use your skill and judgement in transfer dealings, and I have learned the hard way down the years. It is also up to the club to make sure that all are rewarded well for their contribution, if they can afford it. Everyone finds their own level and I know that I am no Dennis Bergkamp but I do know my own worth. Also, there are a lot of people associated with football who are less important than the players but are still making fortunes out of it. So why should the people who produce the product not get the lion's share?

I do not object, either, to quality overseas players coming here and getting large sums in wages. It brightens up our game, from which we all benefit, and if we don't pay the amounts, other leagues will and we will lose the best. I get fed up with people saying what a great league Serie A is. Do they moan over there about the sums players are getting paid? I think the fans just want to see the best and leave it to the clubs to afford them. It is becoming the same over here. We should get as many foreigners as we can.

But as a player who has played in every division of the game, I would like to see a better way of spreading the wealth around, both to clubs and players. I would not have become a Premiership player if I'd not had Aldershot to set me on the way and I hope that the First Division clubs, themselves on Sky these days, do not forget that they used the argument about supplying the Premiership clubs with players when it comes to divvying up the millions that the Nationwide League now gets from TV. The argument did apply to them and it also applies to their lower-division brethren.

The Professional Footballers' Association has also been

criticized for greed when they insisted on their traditional 10 per cent of the TV deals. But for every Premiership player earning a fortune there are at least five pros barely scraping a living. Yes, wages have gone up a lot, but so has the Sky money, and it is only right that the performers get an appropriate cut. Although the PFA has assets of £8 million, they donate a lot back to clubs – medical facilities, community work, for instance – which is another way the game has improved. There is also the benevolent fund to administer to ensure that players get a reasonable sum when they retire.

That, though, is a system that I feel needs overhauling. The lump sum you get at the end of your career works by multiplying by three the number of years you have been a pro – say 10, which makes 30. They then divide that by 80 – I don't know why that figure is used – and multiply the whole lot by the sum of your earnings for the best three years of your career. If that figure is £100,000 per year, that equals £300,000, so the equation is 30 ÷ 80 × £300,000, which equals £112,500.

For a player coming into the game before 1987 there was no ceiling to the lump sum, but one was established then at £100,000 and from the 88–89 season at £82,500, which will be the amount I receive, therefore. My two years dropping down to Weymouth have thus cost me quite a bit. I returned to the full-time ranks in 1988. Had I continued in the league after starting with Bournemouth in 1984, I might have received as much as £250,000, based on my current salary and fifteen years as a full-timer.

Still, £82,500 is not bad and must sound a lot to most people. Most players, however, will never get anything like that. Take a five-year pro at the lower end of the scale whose best three years brought him £30,000 a year. For him the equation is 15 ÷ 80 × £90,000, which means a lump sum of around £17,000. It seems to me that those who are earning the most will get the most, which to me is unfair. For the highest-paid, £82,500 is only a couple of months' wages and they should hardly need it. It is the lower-paid who need better lump sums.

After the Jean-Marc Bosman case we are beginning to get away from the restrictive, feudal hire-and-fire system that has been football. And I think it is all for the good. We are now getting the best stars in this country and our game is as good as it ever has been – I don't care what the old-timers say. Young players can only improve in such an atmosphere and the best will always come through, being even better. It is hard to appreciate just how good the standard in the Premiership is until you play against the best clubs. My first experiences were for Aldershot against Sheffield Wednesday then Cambridge against Manchester United. Their speed of thought is just phenomenal. We couldn't get close to United, with their first touch so sharp and taken at pace. They had half a second and half a yard on you before you knew what was happening, giving them time to pick out the best pass. You would turn round and before you could move, the ball was gone. They had so many options and were so slick, confident and comfortable on the ball.

I couldn't believe the so-called experts who were saying our game had gone downhill, pointing to bad results in Europe in recent years, rectified when Manchester United won the European Cup. When English clubs were winning things before Heysel it was with teams of Scots, Irish and Welsh as well. Then when we came back into Europe after the Heysel ban, we were restricted to three foreigners. It was nothing to do with the state of the game, it was the state of the rules. In the next few years, post-Bosman, I expect us to do well again in Europe. We and the Italians have the best league football in the world and I believe our results in European competition will prove it again. I think England's performances at Euro '96 also showed that we do have quality players in this country.

I felt privileged to be part of the boom but I also felt I had earned it the hard way. Even after my goal for Leicester at Wembley a few of the papers were grudging in their praise, describing me as workmanlike and a journeyman, concluding that I was lucky to be playing at the top level. Well, I may have been a journeyman in many ways, with all the lower division clubs I have been through, but I do believe that there is a lot more to my game. I believe I have a good footballing brain, a decent first touch and good vision. I like best to play up front with either a target man or a natural goalscorer. I am also able to get wide, go past a man and get in a good cross.

Fourteen years after starting out as an apprentice at Portsmouth I had finally made it. I was now playing for another set of fans who appreciated me and the game. Leicester is another of those real football towns where people turn out if they are being served up decent stuff. After Birmingham it had to be another big club. What a thrill just to be playing at the top level and go to grounds I had only dreamed of, to those massive cathedrals that all-seater Premiership stadiums have become. Although Leicester were many people's favourites to go straight back down, I believed we had the makings of a good side. I was sorry to see my new team-mate Iwan Roberts go to Wolves but in Emile Heskey we had a young lad who could go to the very top as a forward. He was strong and quick with real all-round ability. Then there was Neil Lennon, signed from Crewe, who impressed me with his ability to retain possession, and Muzzy Izzet, signed from Chelsea, who had a neat touch. I remember a match against Sheffield United during our promotion drive: the pitch was bone-hard and bumpy, but he came on as a sub and showed how to get the ball down and run with it. He has excellent balance.

Training, conducted by Steve Walford and John Robertson and not being real high-pressure stuff, was right up my street. Steve said I was the worst trainer he had ever seen but the fittest player. Underpinning it all was Martin O'Neill, for whom all the players had great respect given his playing record as a passing midfielder with Nottingham Forest and Northern Ireland – a Championship, two European Cup winners medals and 64 caps – though he insisted he was just a runner and ball winner. He was not an up-and-at-'em sort of manager but had his moments of anger, just to show the players he cared and to make sure they did. He was mostly very studious and didn't panic. Best of all, he didn't try and change your game but played to your strengths.

I remember thinking in the summer, before the fixtures were published, that my play-off final goal would mean us going to Old Trafford while Palace had to go back to Grimsby. Blow me if it wasn't scheduled for the same day, November 30, to emphasize the world of difference a last-minute goal makes. In financial terms, too. It probably meant an extra few million pounds to Leicester, going up, and a shot at the £9 million in TV money alone the following season, so I reckon I had been good value to them. All was set fair, though even I didn't expect the amazing season it was to prove, especially its amazing end.

9 Eat Football, Sleep Football, Win the Coca-Cola

Steve was worried in the autumn of '97 that Premiership football was mellowing him too much. 'I've got really boring,' he told me.

Well, it was true that he had lost no more than £10,000 with his gambling during the previous season, worn the same boots for much of the time, not accumulated any more penalty points on his driving licence and had, mostly, turned up on time for training. But boring? The little matter of a lightning-does-strike-twice goal in the dying minutes of a Coca-Cola Cup Final replay suggests not; as did his Football Writers' Association Midlands Player of the Year Award and third place in *FourFourTwo* magazine's 'Cult Hero of 1997' poll.

Claridge and Leicester City took the Premiership by surprise. They were most people's favourites to go straight back down, but a combination of remarkable resolve and unexpected quality saw them all but safe with several games to go. There was, to boot, the uplifting Coca-Cola Cup win, courtesy of Steve's goal in the replay against Middlesbrough, which earned them a place in the UEFA cup. 'A lot of people thought I was more likely to end up on a park bench than in Europe,' he said.

It was one of 15 goals that he scored in the season, and after that opener – an eye-opener indeed – against Sheffield Wednesday, 11 more followed in the Premiership. It put him in the top 10 of strikers in the top flight – the other nine being Robbie Fowler, Fabrizio Ravanelli, Ian Wright, Alan Shearer, Dwight Yorke, Les Ferdinand, Ole-Gunnar Solskjaer, Matthew Le Tissier and Stan Collymore – and made him the highest-placed striker uncapped by his country.

There was a remarkable range of strike, too. From a predator's tap-in to beat Tottenham at White Hart Lane, through a stunning volley against Manchester United in the Coca-Cola Cup to a mazy solo goal at Blackburn on the last day of the season, Claridge revealed a quality that few had associated with him. It led to the chief executive of the FA Graham Kelly, no less, lauding him in a national newspaper as his 'hero of the week' and a worthy example of the English game's virtues beyond all the overseas stars hogging the limelight.

Of course the highlight was the Coca-Cola Cup Final. In the week leading up to it I travelled to Leicester for an interview with Steve's midfield team-mate Garry Parker for the *Independent on Sunday*. Affection shone through occasional frustration. 'Great bloke, Steve,' he told me, 'but frustrating to play with. He just won't give the ball back sometimes.'

Claridge rarely got the chance to hold on to it at Wembley in the 1–1 draw scrambled with Emile Heskey's goal deep in extra time, as Leicester barely did themselves justice. The replay at Hillsborough ten days later was full of the same forgettable football until Steve struck, steering home a volley after Steve Walsh had knocked down Parker's free-kick. Almost a repeat of the play-off final goal. 'The funny thing is,' said Steve afterwards, 'I didn't back myself to score. If I had done, I would probably have missed it.'

SHOTGUN CLARIDGE, announced the *Sun*. LOVE AND CLARIDGE, said the *Daily Mail*. 'Claridge may not possess socks that stay up but his sense of timing remains impeccable,' wrote David Lacey in the *Guardian*. 'Like both team and club, there is nothing starry about Claridge, simply an honest professional pushing his abilities and appetite to the limit,' wrote Henry Winter in the *Daily Telegraph*. In echoes of my own first interview with Steve, Winter had visited his Luton home earlier in the season, only for Steve to interrupt proceedings by venturing next door to borrow potatoes from a neighbour.

During the season, the Leicester manager Martin O'Neill revealed that there had been opposition in the Filbert Street boardroom to him signing Claridge, such was the reputation that preceded him. 'But I was determined to get him,' said O'Neill. 'I regarded him as a poor man's Kenny Dalglish. I saw him as a vital element in putting together a team capable of promotion . . . and more. You cannot fault his contribution for us.'

O'Neill's first-team coach Steve Walford could, in training, unsurprisingly. 'His attitude with his training was terrible,' he told me. 'Sometimes I felt like slinging him out, like telling him, "You're spoiling it." Yes, I took a few bob off him in fines for late arrivals. We used to start at ten-thirty, and right on the dot Steve would come screeching in to the training ground in his car with his kit on, boots and all. He was getting away with murder. So I made it ten-fifteen in the dressing room. Then he would start turning up at twenty-past, twenty-five-past. I had to fine him.'

Otherwise, Walford was more than pleased with Claridge's input. 'He didn't really surprise us because we knew he had enough ability. In some games he was a different class. Mind you, in others he played like a Third Division player. In fairness, that wasn't just Steve. We had a few like that. He took the goal against Middlesbrough brilliantly. It wasn't an easy chance. He's written himself into the history of the club with that and the play-off goal, hasn't he? But for us, his main contribution was in keeping us in the Premiership. That was the aim.'

Any evidence of the gambling? 'I steered well clear of that,' says Walford, 'but I know he took a beating now and then. I think he ran a book for the lads sometimes.' My own experience of it came when I was driving Steve to an interview with a radio station in West London when, in the middle of traffic, he announced that nature was calling. I stopped and parked on a double-yellow line, expecting him to be back at any minute. After ten minutes I thought I had better go and look for him. Nature had indeed called. As I suspected, there he stood in William Hill, gazing up at the bank of screens.

'He is the strangest bloke I have ever worked with,' added Walford. 'At Torpoint in pre-season training we were on the beach and all the boys were running in and out of the water. He said he couldn't, because the combination of the salt and the sun would bring him out in a rash. He was getting quite panicky but they didn't care. They just ambushed him and slung him in, though he fought like hell. He wasn't happy. To be fair, I think exposure to the sun, coupled with the medication he was on for his heart, does genuinely cause an allergic reaction.'

The hope, for Steve if not City supporters, therefore, was that the UEFA Cup draw did not take him to sunny climes. 'But I won't let a little thing like that stand in my way,' he said. 'I'll just slap on loads of Factor 25 and go out there white as a sheet. It might frighten them.' Actually, although it may be hard to picture anyone being frightened by as genial a character as Steve Claridge, after a season like '96–97 there would be few who don't respect him.

At Wembley in the play-off final, the ball just bounced up and winked at me, waiting to be struck home. This time it fell lazily, almost pleading with me to steer it home on the volley. As before, I knew instantly it was going to find the back of the net but still I could hardly believe it had happened again. In the dying moments of a huge game, I had scored the winning goal. I know football is a team game, but when you are the one who actually scores that decisive goal, the feeling of elation and achievement is that much more special.

That April night at Hillsborough, when Leicester City won the Coca-Cola Cup by beating Middlesbrough, ranks as one of the greatest nights in my career and also the club's history. It produced their first major trophy since 1964. That was for the fans. It may sound unromantic, but for us professionals inside Filbert Street, staying in the Premiership, with some comfort, was the greater achievement.

In the end, many critics put our success down to team spirit and hard work. But that explanation just got them off the hook. I think our quality was underestimated. While we may have possessed those attributes, the team added up to much more than that. You don't accumulate points in the Premiership or win a trophy just on workrate and determination.

It is hard to convey just what an intense experience the Premiership is. All the games are massive to a club like Leicester, which is the main difference from the First Division, where we were often the big fish that other teams wanted to hook. Perhaps it is not the same for Liverpool or Manchester United – they probably know that they are going to stay up no matter what, but for us every point gained was a triumph. We knew from the outset that 42 points and safety was our target, and every time you took a step towards that aim you really thought you were making progress.

That was probably what made the season so exciting and so enjoyable. Every game meant something, with an end-of-season edge to each one. Our consistency was the crucial factor. We never had blips, losing six games in a row, for example, except for a slightly iffy period around the Coca-Cola Semi-Final and Final. Even then, though we went nine games without a win, we still drew five of them. When it did look dodgy for a brief spell – only two places above the trapdoor with only two matches left – we promptly went out and won both games to finish a more than creditable 9th.

I don't think the Premiership was harder or easier than the First Division, just different. The tactics are much the same. The main thing is that there is a greater depth of quality in each club. In the First Division every team has two or three good players, but if they are injured or suspended the overall quality of the team suffers. It's not the same in the Premiership. Players who come in to the team are top quality and consequently you never get an easy game.

Neither is the game necessarily any quicker in the top flight, except in one crucial area – speed of thought. Players read the game far more sharply, defenders nip in ahead of you to nick the ball much more often. You don't get chopped or buffeted as much in the Premiership, so physically I actually found it easier, but mentally I always left the field drained because concentration levels were so much higher for that much longer in a match.

Sometimes, before I played in the Premiership, I used to watch all these players on the telly and think that I fancied my chances against them. They just didn't look that good. But when you are on the pitch with them, you realize that they are. It changed for me, after quite a few tussles with central defenders, from 'I wouldn't mind playing against him' to 'I wouldn't want to play against him every week.'

The best defence *en bloc* was Arsenal's. People were telling me they were getting on, but for me they remained the masters. So well organized as a unit. After we lost 2–0 to them, I backed them at 16–1 for the title. Individually, I thought David May at Manchester United and Colin Hendry at Blackburn stood out. At Middlesbrough Gianluca Festa was all you would expect of an Italian: rugged and disciplined. He was all over you like a cheap suit. I should know, I've worn a few of them in my time.

I do think the standard of defending in the Premiership has improved a lot. You are even seeing teams like Chelsea, who were renowned for having a soft spot in the past, defending solidly. Their FA Cup win of that season may have owed a lot to Gianfranco Zola up front, but it was based on Frank Leboeuf at the back. They should never have got past us in the fifth round, mind, because no one will ever convince me that Matt Elliott brought down Erland Johnsen in our replay with them at Stamford Bridge – though I have to admit I only saw the incident on TV in the players' bar as I'd left the field early with my arm in a sling, having dislocated a shoulder by an inch and a half – but that's another story altogether. It left a nasty taste in our mouths – who knows, perhaps it did us good, strengthening our determination for the Coca-Cola.

As for strikers, the one who impressed me most was Dennis Bergkamp. We thought he might be a bit lightweight but he was much stronger than we believed, as our Captain Steve Walsh, who knows a bit about tough players, said after our match against Arsenal. Bergkamp can certainly put himself about and his touch on the ball was superb. There were many times when I thought he dragged Arsenal through games almost single-handedly, as Eric Cantona had done for Manchester United the year before.

And you have to be strong in our game. All the big names coming from overseas have first got to show they can hack it physically. Those who have left quickly have, by and large, not been able to. It's not quite the easy money they thought it was sometimes. People go on about our league being all about excitement rather than quality, but that's not how I see it. It is a blend of the physical and the skilful, the fast and the entertaining, that still makes it the best in the world.

As for my own game, what pleased me most was the rate at which I took my chances. It wasn't that I was getting six or seven a game. I probably only had thirty to thirty-five chances over the whole season, and converted fifteen of them. I don't think I have ever scored that ratio of goals to chances in my whole career. At Liverpool, for example, I was allowed only one shot at goal and scored with it to earn us a point. What a thrill that was, scoring at Anfield, where I used to travel up from Portsmouth as a kid. I was also pleased at my durability. I did pick up a few injuries – an ankle at Tottenham, the shoulder at Chelsea – but they were impact injuries rather than anything to do with ageing limbs. After the shoulder injury I even played a few times with it padded up. Out of fifty-one competitive matches, I missed only seven.

Several goals stand out in the league, like the one at Liverpool (for sentimental reasons), the long-range shot against Sheffield Wednesday, and a solo effort on the last day of the season at Blackburn, where their defence just opened up and I showed a turn of pace not many thought I had; but the most memorable just had to be in the Coca-Cola.

The campaign began poorly for me – as per usual in these things. In our opening game in the second round at Scarborough I had a nightmare. They played five at the back, four in midfield, and every time I got the ball men were swarming around me. Mark Robins and I couldn't move up front. Bloody awful it was, and I was substituted, but the boys won comfortably enough: 2–0. I also missed the second leg, which we won 2–1, having injured an ankle in scoring a winning goal at Tottenham the previous weekend. Next we went to York and also won 2–0, which was a better result than it might look as they had knocked out Everton in the previous round. They had also beaten Manchester United over two legs the previous season. Now it earned us United at home in the fourth round.

It has to be admitted that for Manchester United, the Coca-Cola Cup is their lowest priority of any season these days, and they sent a weakened side to Filbert Street that night. A weakened side for United, though, is still good enough to do well in the

Premiership. They had plenty of players with first-team experience and still gave us plenty of trouble. We were fortunate that we got a goal at the right time, just before half-time, and that they also missed a penalty through Paul Scholes.

What a goal it was. Simon Grayson took a throw-in on the right, Garry Parker half-volleyed it into the area, Emile Heskey flicked it back across goal with his heel and I spun off my marker to volley it home on the run. It was an amazing move and had anyone else scored it, any bigger team, the acclaim would surely have been greater. When Emile sealed another 2–0 win in the second half, it dawned on us, for the first time I think, that we could win this competition. The bigger teams were all falling by the wayside and we were in the quarter-finals.

The feeling was confirmed when we won 1–0 at Ipswich, a game which saw me awarded a mountain bike as man of the match. Very handy it has proved too, cycling up and down the bumpy, unmade road that runs past my parents' place near Portsmouth. It was also a win that showed how far we had come. Ipswich were one of the best teams in the First Division, and might have gone on to win the play-offs as they looked the classiest side in them, but we were never in any danger, even if the scoreline did not reflect our superiority. That night, in the course of which they only really had one chance, there did seem a big gap between the divisions.

It paired us against Wimbledon in the Semi-Finals, and what a grind the two legs proved to be. We thought they had changed the way they played but it certainly wasn't the case in those games. They still lumped the ball forward at the earliest opportunity – it was like playing against Cambridge United in the bad old days. They were daft, really, because in people like Oyvind Leonhardsen and Marcus Gayle, they had some good players.

They would have been better off trying to play football against us. We could soak up all the aerial pressure with three centre halves at the back: Steve Walsh, who always loves a battle, Julian Watts and Spencer Prior.

Many people thought our chance had gone after we drew 0–0 in a home leg we should have won. Emile hit a post and I dribbled one shot just wide after I was clean through. Still, we felt we could do it. As is proved in Europe these days, as long as you don't concede at home you have a great chance of going through by pinching the away goal. We had done it at Stoke in the play-offs the previous season.

It was a real battle at Selhurst Park and in the first half they battered us, scoring a typical scrambled goal through Gayle. But in the second we began to put our game together after absorbing their pressure. Our equalizer came through an unlikely source in Simon Grayson. After that, we held on through extra time, with Garry Parker twice clearing off the line, to go through. It wasn't a classic performance but we were at Wembley.

So was something very unusual – a ten-foot-high effigy of me. Actually, it was a huge head on top of some poor bloke sweating underneath it, as part of the pre-match entertainment. Leicester were also represented by one of Martin O'Neill, while Middlesbrough had Fabrizio Ravanelli and Emerson. I think mine was the slowest in that as well. A couple of weeks later I tried to track it down, to try and buy it, but without success. It might have looked good as a landmark on the roof of the new house I was building next door to my parents.

I think we were too negative in that game, playing with only me up front and Emile wide on the left, and defending too deep. We got too stretched out and whenever I got the ball I didn't feel I had enough support. I think the gaffer was worried about the way Juninho had taken us apart in the league. This time Pontus Kaamark was told to follow the Brazilian about, to great effect as it turned out, even if Pontus did have reservations about the job, worrying that it may not be in the spirit of the game. My own view is that in the professional game, it is mostly about what gets the job done.

After a goalless ninety minutes, they grabbed a lead through Ravanelli five minutes into extra time, and it seemed fair enough; if anyone was going to win the game that day, it was going to be them. But our fighting qualities shone through again, our ability to grind out a result when not playing well, and as soon as Emile bundled home an equalizer with a few minutes to go I felt that the momentum was now with us and that we would win the replay.

That, too, was a mostly forgettable match, except for its finale. It was incredible how the circumstances were virtually the same as those against Crystal Palace in the play-off final. Garry Parker took a free-kick, Steve Walsh knocked it down, and there I was to grab the dramatic winner. As it was falling, I was just concentrating on hitting it sweetly. I knew where I wanted to put it – low to Ben Roberts's left. And lo, that's where it ended up. I didn't try to whack it, or steer it and control it, because I knew that if I caught it flush from such close range there was no way he was going to stop it. It was yet another example of taking your chance in life when it comes along. No matter how much I earn or don't, or whatever I go on to do in life, I will always have that memory now.

The goal earned me quite a bit of celebrity, and out of all the publicity something strange happened. One day soon afterwards I received a phone call from a young bloke who had read that I was adopted. He believed he might be my half-brother.

When I met up with him, somewhat nervously, his story sounded feasible, with dates and the circumstances of birth sounding similar. There was no doubt he was genuine and not a hoaxer. He was a really nice lad. But my mum Anne looked into it further, even tracking down my blood mother on the Isle of Wight, and it proved not to add up, to the lad's disappointment. It stirred something in me, though, and I began to feel, slowly, that I wanted to find out more about my origins.

The goal did also earn me a few bob. In the summer, Martin O'Neill acknowledged my request for an improved contract and my basic went up to £4500 a week, with bonuses taking it near to £5000. I felt I was worth it. Some new signings had come in on more than me and I was the leading goalscorer, so I felt that at 31, with time catching up, I had to make what I could while I could. It was good money, I accept, even if it didn't compare with what the top players elsewhere were on.

Martin was a very shrewd man. He knew what he wanted and bought players very well, as was shown in the case of Matt Elliott, for £1.5 million from Oxford United. Matt proved an excellent signing, a much better footballer than people gave him credit for. He was much more than just a big, strong stopper, though he was not as quick as he thought. He kept saying he was going to beat me in a race to the halfway line, so I was keen to have a few bob on that. He was typical of the type of player Martin looked for, one with something to prove, in their mid- to late twenties, players who hadn't achieved what they could and were willing to come and work. The gaffer had the knack, a bit like his old Nottingham Forest manager Brian Clough, of buying the right player at the right time and for the right price. He deserved his own improved contract in the summer and the players were all glad to see he wasn't tempted away by a bigger club.

Martin's main criticism of me was that I didn't seem to be around the place enough. Well, I am a bit vague, aren't I? I think the lads would have agreed. A few have said that they would like to see more of me – even if some of them wouldn't – but I guess I can be a loner. They were a good crowd, who socialized and enjoyed the banter together. There wasn't one of them I didn't feel I could talk to when I walked into the dressing room.

It was a mixed crowd. You had a quiet, studious guy like Kasey Keller, who kept himself to himself (and who was quoted during the season as saying that we were not the most intelligent bunch, which I suppose was pretty accurate), as well as family men and more happy-go-lucky characters like me. Pontus Kaamark was also quoted back home in Sweden as saying that he couldn't believe how fond of the booze English players were, and that some had women in their hotel rooms the night before a game. It caused a bit of a stir in Leicester and Martin O'Neill banned him from talking to the press for a few weeks, but mostly the lads just laughed it off. We were like any other group of young men, who like a drink, a bet or a woman's company, but I never saw it happening to excess at Leicester. All of us liked a good night out, particularly the livelier single guys, like Muzzy Izzet and Neil Lennon, but never before a match.

Probably the funniest guy in the team was Garry Parker, who called me the Morocco Mole, after some cartoon character, because I kept turning and turning when I had the ball. The thing was, I don't like giving it back to him too often because he will just whack it forty yards and give it away. It was not an easy season for Parks, with a worrying time when his wife delivered a baby after only twenty-three weeks, which thankfully survived, but he bore it all very stoically and still kept his sense of humour around the club.

One day the team were having massages at a local college. When it was my turn on the table, Parks and Scott Taylor came in to talk to me. They were being unusually friendly, I thought, and I soon found out why. After my massage, I discovered that my clothes were missing. I then had to run all the way to the reception area of the college in just my underpants, where the rest of the team were in fits of laughter, Parks and Scott clinging on to my kit. It was almost as embarrassing as the day we all went swimming at a local baths. At one point, Steve Walsh dived in and, obviously with his eyes shut, tried to grab one of the lads' legs under the water. It turned out to be some poor unsuspecting old lady who came up coughing and spluttering.

On the timekeeping front I was not too bad, by and large. I did hit a purple patch when I was late for training quite often, and Steve Walford introduced a system of fines – a pound for every minute you were late. I'm sure I managed to keep him and Martin's assistant John Robertson in beer and fags for the season. I think I averaged about five pounds a day and it got to the stage where I had to keep some cash in the ashtray of my car.

They often used to say that I was taking the piss, but I reckon it was the other way around at times. When they were short of cash they would fine me for anything and everything – once for walking down a corridor the wrong way, I think. They were lying in wait for me every day, and while I was being lectured other latecomers would get off scot-free by sneaking round behind them. Sometimes I was late for training because I had to drive so slowly. I couldn't risk speeding as I was just one point away from a driving ban, and I just couldn't be without a car. I was stopped a couple of times for a rear brake light but got away with it. Once, I heard a siren behind me and, thinking my number was up, slowed right down. It turned out to be some music on the stereo that featured a siren.

With the gambling, it was a season of ups and downs. I had a good win on us lifting the Coca-Cola Cup, taking £1000 at 3–1, and another when we stayed up, £2000 at 7–2. Another private bet I struck on Manchester United to win the title I agreed to cancel when I received a letter from the FA, who read about it in the papers and pointed out that we had still to play them, with me in a position to influence the outcome.

I was becoming more disciplined about my betting now. If I like the idea of something, I went for it. It was amusement, really. I probably lost about ten grand over the season, which wasn't bad for me. Having my house built helped, with all the costs involved, because it stopped me spending more. I suppose buying property was my roundabout way of stopping myself wasting it. If I just deposited it in the bank, the temptation to get at it would have been too much. My biggest loss was actually a win. A bookmaker owed me £37,000, but he went bankrupt. I was offered his shop as payment, but I could see only new problems in that. Taking on closed premises with all the customers having drifted elsewhere was probably the least of them.

The gambling being under control, and everything going well for me at Leicester with a UEFA Cup campaign to look forward to, meant that life was good. Just when you thought you had it all sort, though. . . .

10 Back Home

Nothing much seems to come as a shock or setback to Steve Claridge. Tommy Docherty once said that in football, one door closes and another one slams in your face but Steve has always looked for the next door ajar, packed up his boots and his troubles in his old kit bag and smiled his way to another club whenever things have run their course or gone wrong.

He had indeed become a legend at Leicester with those two goals, to get them into the Premiership then win them the Coca-Cola Cup, but for Martin O'Neill, legends had to be living. Past achievements counted for nothing when Steve returned for the 1997–98 season and he found himself out of the side as the manager looked to a new striking partnership. In America they call Major League Baseball 'The Show'. In this country it is what the Premiership has become but Steve felt that for him the show was over.

So it was that he believed he had to move on as the season unfolded. These days squad rotation may be all the rage at the biggest clubs who have European competition as well as the Premiership to occupy them, but Steve needs to be playing week in, week out for his own contentment. For him, no amount of money makes up for being a substitute and just making the odd appearance. 'Go where they want you, at whatever level,' has always been his motto.

Returning for a loan spell to his boyhood home-town club of Portsmouth was a blast and he wanted to stay after three months with the salt of the Solent in his nostrils again. At first he admired Alan Ball, back for a second spell as manager, for admitting he had been wrong to release him all those years ago and for trying to recapture the pride in the town and area which Steve himself felt. The club was in terrible financial straits, however, and could not afford the £400,000 transfer fee Leicester wanted. However, Wolverhampton Wanderers could and so on deadline day came a move to Molineux.

It was an unhappy period, one to rival the few months at Luton earlier in his career. Wolves were in an FA Cup semi-final against Arsenal, but never looked like halting their march to the Double. They were also close to the First Division play-offs, but not close enough and their challenge petered out. The challenge for Steve of replacing another cult hero in Steve Bull, meanwhile, was too onerous and with Mark McGhee nearing the sack, Wolves let him go back to Pompey, the price now having been reduced to a more manageable £175,000 for them.

He was delighted to have come home and settle into his newly built house. Before too long the relationship with Alan Ball had soured, however, and he lost respect for the manager amid arguments and declining results that threatened another relegation struggle. He was dropped and considered a loan move to Brighton and Hove Albion. He was reluctant to move on, though, having worked so hard to get back to where it all started.

In the end, the feeling from within the club's heirarchy and from fans on the terraces, who saw in Steve Claridge one of their own, was greater for him than for Alan Ball and it was the manager who departed. Steve's form picked up and he began scoring regularly again, winning the Player of the Season award two years in succession.

He has mellowed with age and experience, he insists, but he greatly exaggerates it. 'Talking to people around the club he has mellowed from when he was a kid here,' says the Portsmouth captain Andy Awford, 'but he still lives life on the edge. It seems upside down sometimes, just like that old kitbag. His whole life seems to be in there. Sometimes there's wads of money in there as well as all sorts of boots and shin pads.'

Ah, his shin pads. He still has a favourite old pair. Not a pair exactly, but two odd ones. He is most particular about wearing them, as a story told to me by his mother Anne illustrates (Steve was late for our appointment and she had kindly made me a cup of tea). It is a tale that also illustrates the still at-times chaotic but endearing nature of his existence.

It was 2.50 p.m. on a Saturday afternoon and Portsmouth were at home to Swindon. The phone rang in the Claridge household. It was Steve from the dressing room to ask his mum to pick up the shin pads he had forgotten from his house next door and to drive to Fratton Park with them. Mrs Claridge admits to not always being the most confident of drivers; she would dig them out, she said, but suggested that he send a taxi to pick them up.

At 3.35 p.m. the hapless taxi driver arrived, flustered after negotiating the country lanes to find the place. He sped off with the precious cargo on the front seat, and ran the final hundred yards to reach the players' tunnel just as Steve was leaving the dressing room to start the second half. He then had to wait for his fare until the final whistle.

'He is disorganised,' adds Andy Awford. 'He is always the last in to training. But he always comes up smelling of roses, doesn't he?'

My own experience of that came shortly after Steve had signed for Wolves. They were at Queen's Park Rangers in midweek and I arranged to watch the game before going for a meal with him. Half way through our curry, he suddenly remembered that he had left his pay-off cheque from Leicester – all £48,000 of it – in his room at Wolves' pre-match hotel, the Royal Lancaster. There followed a midnight dash back to West London, the cheque fortunately having been handed in to the concierge, before I took him to Toddington services on the M1 where he had left his car.

Even though he was delighted to be back at Pompey, it was always probable that there would be turbulence given the passionate characters of both him and Alan Ball. 'Bally loves a rogue,' says Andy Awford. 'I think he saw them as a bit of a challenge and we have got plenty of them at Portsmouth. Leaving crossed Steve's mind when he was clashing with Bally, I'm sure.'

One of the final insults in Steve's eyes was that Ball was getting good tips for horses but not telling him! Steve being Steve, he got them from one of the other lads. His gambling, mostly, has been under control, though the FA took a dim view of one bet he had on Pompey to beat Barnsley, when he scored a hat-trick. He was summoned to a disciplinary hearing. As he admits, there was a reason why he himself revealed the bet publicly – a reason which in the end backfired.

It was always likely that Steve's much-underestimated talent would prevail at Portsmouth despite the dust-ups with Ball, along with his unquestionable work ethic. 'He's so wholehearted on the pitch,' says Awford. 'The players and the fans always know that he is giving his all and I think they see a lot of themselves out there at Fratton. They know he feels the same as they do about the game and the club.

'There is a good understanding between him and the rest of the team. As a defender, you know he likes the ball into his feet but you know as well that he will get after the long ball even if he doesn't like it. Opposing defenders may not worry about the ball over the top for him to chase because he's not the quickest any more, but they are right to be concerned about him when he's got his back to goal because if they get too close, they know he can do them with his little turns.'

There is rivalry, too, between Claridge and Awford. Claridge has a column midweek in the local paper, *The News*; Awford has one in the Saturday *Sports Mail*. 'He kept going on in his about how big my arse was,' says Awford. 'Then he was unhappy when I told the lads to back a horse I have a share in and it came in next to last. So I didn't tell him the next time it was running, when it had a better chance and, in fact, won. He was not happy about that. I had got my own back.'

Awford should not bank on that being the end of the matter. If Steve Claridge's career has proved anything, it is that he has had plenty of laughs with, no doubt, more to come. And he usually has the last one. One day in the not too distant future, management will beckon. I've told him that when I win the lottery and buy Weymouth FC, he has the job.

There are no guarantees in football, my whole career had told me that. You just live from one season to the next, hoping that it will go well, fearing that it could go wrong. I wasn't really prepared for what happened to me just a few months after that winning goal for Leicester in the Coca-Cola final, though.

Martin O'Neill seemed fine with it when I opted out of the club's pre-season trip to Greece. The way I am with the sun, I only go abroad if I have to. I hadn't gone the previous season and it didn't seem to matter. This time it must have, though. The manager started the season with Emile Heskey and Ian Marshall up front, the team won four and drew two of their opening seven games, and although I got a couple of starts I felt very much out of favour.

Even in those two games, I was pulled off and I definitely got the hump. In the away leg of the UEFA Cup tie against Atletico Madrid, I got on as a sub for Marshall after about half an hour but was then subbed myself in the second half. What upset me most was that it was Marshy ahead of me. Over the last couple of years, I had played with loads of injuries, a broken rib and everything, but he had to be 110 per cent fit to play. I liked him as a bloke but he annoyed me as a player. I had run my socks off pre-season and it had seemed like a joke to him.

I did have a couple of rows with Martin O'Neill those times he gave me the hook – players' jargon for being subbed – and we did fall out over it. I was daft, really. I should have just bided my time, because I would undoubtedly have got back in. Martin was always straight with me and told me so. But when you're in those positions, sometimes you can't see the wood for the trees. By Christmas, I wanted to get away and I was beginning to feel the urge to get back nearer home. I had rung Portsmouth a couple of times before when I was between clubs. When they heard I might be available now, their caretaker-manager Keith Waldron – Terry Venables and Terry Fenwick having just departed the club – contacted me and said they couldn't afford to buy me but would take me on loan.

I was delighted to be back playing first-team football again, but Pompey were in a sorry mess. Martin Gregory was in charge as chairman and, despite the best efforts of Terry Brady, who I liked as general manager, the shambles was reflected on the pitch. In fact, at that time, I think Pompey were probably the most badly organised team I have ever played for. They were bottom of the First Division by some distance. It was only a passionate group of fans that seemed to be keeping the club going.

As usual, I had a terrible start with them as well. My first game took us to Oxford and I can't remember ever doing well there. We duly lost 1–0. Then it was a home game, marred by an assault on a linesman. It was against Sheffield United and we had gone a goal up before our goalkeeper, Allan Knight, caught a cross, fell backwards, banged his head on a post and fell into the net. If we hadn't been in such trouble, we probably would have laughed about it. We did afterwards.

Before then, Sammy Igoe was racing clear and the United goalkeeper took him about 20 yards outside his own box. I saw the linesman, a Mr Edward Martin, flag for the foul, and draw it to the referee's attention that the keeper should be sent off. The next thing I know is the lino is spark out on the ground. Some lunatic, a Sheffield United supporter living in Sussex, was later convicted of the assault. Poor Mr Martin was out for ages. Yet again, some major incident happened around me. Wherever I go, it does.

My first goal for my home-town club came against Stockport County. I remember the night vividly. Being back at Portsmouth was all the more memorable because the place had not changed one bit since my schoolboy days on the terraces. Even the crush barriers I had painted as an apprentice were still there. The memories were wonderful.

And that night against Stockport I could feel the atmosphere from the terraces again. It was a match we had to win if we were to start our climb up from the bottom and, although there were only just over 8,000 in the ground, they never stopped singing the Pompey chimes – 'Play up Pompey, Pompey play up' – for 90 minutes. Supporters still talk about what a great night it was. I got the ball, played it out wide to Sammy Igoe and made a run, then got in front of a defender to head it home from six yards. It was one of the highlights of my career. I loved it and the fans loved it.

It marked the end of my loan spell and I heard from Leicester that Charlton Athletic had made a move for me. Professionally, a loan plus a transfer if they went up to the Premiership, made sense. They were bombing at the time and were offering the same money as Leicester – about £5,000 a week and £500 appearance money – plus a bonus if they went up. Pompey were offering me nothing. But the pull of home was strong and it was time for me to have a solid base in my life. Alan Ball had also returned as manager to the club he had once taken up into the old First Division and maybe we could turn things around. I signed on for two more loan months.

At first, everything was great with Bally. He took us for coaching and I reckon it was the best coaching I have ever had, with the variety and timing of his sessions. You never felt you were out there too long, just enough to get a sweat on. I have to say, even then I never felt he was going to be a good manager or tactician, though. His first talk to the lads impressed me. He spoke of the pride of the town, that the Royal Navy set out from here to win the War. There was a great tradition and history to the place and the club, which I knew only too well and he captured the spirit as he tried to enthuse the lads. Then he turned to me and said that he had made a mistake in letting me go all those years ago and didn't intend to make another. I was impressed at the time.

It didn't start well, with a 3–1 defeat at Crewe, but we began to get our act together. We won our next four games and soon had moved out of the bottom three. Even then, though, my doubts about Bally were surfacing. Sometimes he would announce his side just before kick-off, with players being played out of position, when you hadn't worked on anything with them during the week. To be fair, he only had 14 or 15 players, but then again we probably did reasonably well because we got a settled side as he couldn't change things too much. I was usually paired up front with John Aloisi, who I thought was a great player.

When my loan spell was up – the maximum permitted is three months – I still wanted to stay but Portsmouth could not afford the fee to sign me full-time.

So I went back to Leicester and, due to an injury crisis, played one game at Wimbledon. I was never going to be flavour of the month again, though, even with the fans singing my name that day at Selhurst and, when transfer deadline day came, I was off again. I still wanted to go back to Fratton Park, and we gave them until 3.45p.m. to stump up the £400,000; Leicester were even willing to take it in three instalments, but they didn't bite. Wolverhampton Wanderers did. They matched my criteria for a move – a club with potential and a passionate following. Under Mark McGhee, they had also reached the semi-finals of the FA Cup and would play Arsenal.

Before then, my first game was against . . . Portsmouth, at Molineux. The Pompey lads stayed in a hotel at Sutton Coldfield on the Friday night and I even went and had a meal with them. The next afternoon, I got a bigger reception from the away fans than from my new club's. The problem was that, in their eyes, I was replacing Steve Bull, who had just been dropped. It is hard following a legend. It pained me that we beat Pompey, but duty called. Our next game was in midweek at Queen's Park Rangers, notable for a buffeting I got from Neil Ruddock and Vinnie Jones.

Then came the semi-final at Villa Park. We were a mess and, after Christopher Wreh had given Arsenal an early lead, we never looked like getting back in it. McGhee put out a formation of 5-2-3, with the full-backs supposed to push on, but they were just running into our wide players, Don Goodman and Paul Simpson. Patrick Vieira and Emmanuel Petit did for us in midfield. That was their strength and we didn't combat it. They were almost taking the mickey out of us.

In theory, we had an outside chance of the play-offs, but I never got the impression that we would make it. By now I had discovered that McGhee had lost the plot and the fans were calling for him to go. He had made several signings just before the deadline and we were all looked on as panic buys. I played up front with several partners – Bully, Goodman, Mixu Paatelainen, as well as Robbie Keane, who even then, at 17, looked the bee's knees. Real bright spark on the pitch with bags of skill and pace. But we just played so many different systems, there was no consistency. It had gone. The whole club was wrong.

I didn't enjoy it and I was nowhere near my best. I think, given another year, I probably could have shown them what I was capable of but it wasn't going to happen under the current circumstances. I was even left out for a game and it bothered me that I wasn't as upset as I should have been. It had always hurt like hell before. Mark McGhee could see it wasn't working out and he was, as usual, straight and honest with me. He said he would sell me at the end of the season.

I waited all summer for another call from Pompey, who knew I was available again. I kept reading in the local paper back home that they were giving trials to all sorts of strikers and I wondered if Barry Fry had taken over as manager. I was on the market for £200,000 and they even went and bought Nicos Kyzeridis from some Greek club for £100,000. I felt like ringing up and saying 'What about me?' Even the local paper was wondering why they weren't going for me.

July came and nothing happened, so I went on a pre-season

tour with Wolves to Austria and Germany. McGhee called a team talk and said that the slate was going to be wiped clean and that this was a new start. I think we were all surprised that he was still in a job. It isn't so easy to turn it round just with words, though.

Personally, I enjoyed the tour, even though we were often bored rigid during the two weeks. We were in nice hotels, but in the middle of nowhere and at one there was only one games room as entertainment. There was no telly we could understand. It was ridiculous. Everyone would be fighting over the table tennis and, if someone got the hump because they were not playing, they would just stamp on the ping-pong ball. There were also a few bats broken. These football hooligans abroad, eh?

By now I had got used to the idea that I would be staying with Wolves. I was happier there than I had been for a while and thought that I would give it a go. Then, finally, in early August, Portsmouth came in and McGhee let me go for £175,000. I think they had tried everything and everybody, and I was the last resort. I wasn't eligible for the opening game, a 2–1 defeat by Watford, but became a fully fledged Pompey player in a midweek Worthington Cup tie against Plymouth Argyle. Home at last. Mind you, it continued my sequence of bad debuts. I came on for the last 20 minutes and missed three one-on-ones with the Plymouth keeper. Their fans were chanting 'What a waste of money.' At £175,000? Still, I suppose that is a lot of money for them. In our dressing room afterwards the lads were wondering if I was the same player that had been on loan the previous season. Up in the directors' box, the Pompey officials were probably wondering if they could still cancel the contract.

I definitely got the feeling that Bally didn't really want me back. I think it was Terry Brady who suggested it. At one point in the negotiations I was told that if I didn't sign that day, the deal was off. I had to take a pay cut, my money going down to around £4,000 a week, but at least I got a longer contract than at Wolves, three years to their two.

How we survived that season, on and off the field, I will never know. A lot of it was down to the enthusiasm of the fans. The debts were running into millions, but we were still signing players, like Thomas Thogersen from Brondby, on four or five times what they were getting at home.

My misgivings surfaced after the first few months over Alan Ball's treatment of John Aloisi. We were working well together up front, me laying on a lot of the 17 league and cup goals he scored in the first half of the season. But he was getting pulled off regularly. It was a joke. The manager would later say that he wouldn't want to be in the trenches with John, but how many games of football are played in a trench? He may have been a laid-back character, but not in front of goal, he wasn't. Pompey made the usual noises about needing to sell him to pay off some debts but, of the £650,000 they got from Coventry City, they were only left with about £150,000 after sell-on fees back to his former club and other deductions. It was like pouring a bottle of water on to a desert. Coventry had even offered £1.5 million a while back which made the whole transfer seem very strange.

The chairman even announced one day that we were all available for transfer. So in our next home game, the lads all had T-shirts made with prices on them – '£2.50 and a sack of spuds' was typical – to wear under our shirts. Mine said: 'Free £10 bet with Ladbroke's'. When we scored, we all lifted our shirts to show them off.

It went from bad to worse. We had reached mid-table in the autumn, but now it was a relegation battle. We had right-backs playing left-back, no wingers, no shape and no formation. I don't think Bally got much help from his assistant Kevin Bond, the former Norwich and Manchester City player, who, in my opinion, had a personality which reflected his game when he was a player – little character.

There was one time at training when we were practising a simple free-kick routine. One player would run over the ball and another would chip it in. Bondy wanted the defenders to charge the free-kick as soon as the first player ran over the ball, but we pointed out that this would be encroachment and we were likely to get booked. He told us not to worry about it, just to do it. We told him that the first player had to touch the ball before the defence could charge, but he wouldn't see it. The argument went on for a few minutes with the lads laughing themselves silly. In the end the manager had to come over and sort it out. You couldn't have too much respect for a bloke like that.

Robbie Pethick also fell out with Bally and was dropped. So Bally then played the left-footed Matty Robinson at right-back. Then again, he could be too soft on some players. Take David Hillier. I thought he was a great player, even if Arsenal did let him go. There was no fear in him. But somehow he was often injured. I would have had him in the club every afternoon working on his fitness, so that we could get the 40 games out of him we needed.

But I was happy enough and playing well, enjoying taking on more responsibility for scoring goals now that John Aloisi had gone. Mostly I was playing with John Durnin up front, though I had games with Stefan Miglioranzi, who was an American-

Brazilian. Actually, to be honest, he played more like an American than a Brazilian. Just a joke – honest!

Somehow we struggled on, picking up the odd win here and there, my goal against Huddersfield deciding it, for example, and we just kept our heads above the relegation line. In fact, we were safe just before the end of the season, even though we lost our last two games and went through four consecutive defeats in March and April. I finished with nine league goals for the season and my performances earned me Player of the Year from the fans, which really thrilled me. I was one of them, after all, even if I didn't get a vote.

The gap between us and the Premiership was huge, though. We went out to Wimbledon in the Worthington Cup, losing 4–1, and to Leeds in the FA Cup, beaten 5–1 at home, though I was personally pleased with my own performance. Despite that, I was really optimistic when the new owner, Milan Mandaric, came in, Martin Gregory departing as chairman. Alan Ball was given money to spend and he got through about £4 million of it in the summer on the likes of Jason Cundy from Ipswich and Rory Allen from Tottenham. My old mate at Wolves, Steve Sedgley, had warned me about Jason Cundy. Nice lad, he said, but a devil when he's had a few to drink. I didn't believe it at first. Jason was such a decent, clever lad. Then one night we all went out to a nightclub. Jason began slide-tackling the lads on the dance floor, catching Buster Phillips and sending him sprawling. I was grateful for Steve's warning. I managed to avoid the tackle that came my way.

Mind you, with Steve Sedgley, it was a case of the pot calling the kettle black. He told a story about when he used to be at Tottenham and every morning for three weeks at training he would throw the goalkeeper Erik Thortsvedt's underpants out of the window. Erik never knew where they were or who had done it. From wearing expensive Calvin Klein's, he ended up coming into training wearing ones from Millett's or somewhere like that. When Erik's patience finally ran out, Steve owned up, taking him outside where there were 20 pairs of pants hanging from a tree.

Anyway, I always said that if Bally had 30 men in a battle against 30,000, he would really fire them up to give a good account of themselves. Give him the 30,000, though, and he would be lost. Now he had the resources, but couldn't make them tell.

In the summer I finally had a knee operation that I had put off. I had sustained the injury at Leicester when I got a kick that pushed bone into the tendon and needed shaving under the patella. It was a big op, the one that finished Rob Jones at West Ham, but I wasn't too worried. I always felt it would take time and it did.

We went down to Nigel Mansell's country club in Devon for pre-season and I could see that the work they were doing was insufficient. On my own in the gym, I was working harder than them on the bike and running after two weeks unable to move. Never once did the manager speak to me during that time to offer any encouragement. Then, when I was finally fit, he wasn't going to play me. I had to go through six reserve games. I wouldn't have minded so much except that when his new signing Rory Allen was injured, he went round with his arms around him. And Rory only had to play one reserve game before he got back in. That, to me, is poor management. You can't have different attitudes to players. Such unfairness creates a divided dressing room.

Even after a 3–0 defeat at Blackburn, Bally wouldn't pick me, the club's player of the year. Instead I was in the reserves at Leyton Orient. When I did get back in, due to injuries, Kevin Bond said out loud that he thought we hadn't had a decent strike force since Rory and Luke Nightingale had been injured. I thought, 'What am I here? Chopped liver?' Rory had just come to the club and Luke was 19. This was just before I was about to play again.

I got back in about six weeks after I should have. Bally had run out of options. We lost 4–0 at Crystal Palace and he went into one of his outbursts where he put half the team on the transfer list. Now the local paper and supporters were asking why I wasn't in the team. I was getting the hump with him but I resolved to stay calm. I hoped he wasn't going to be at the club as long as I would be. He really had to put me in against Charlton and, although we lost 2–0 to a team running away with the league, I thought I played well. I did well, too, I reckoned, in the 5–1 win over Walsall in the next game and scored twice, but got pulled off. I would like to have had the chance to score a hat-trick. It felt like Bally was trying to wind me up. He succeeded and I had a go at him. Now I was getting pulled off every game. He did it at Manchester City, where I thought I was one of the best players, and felt it was getting personal and not in the club's best interests.

The game against Fulham away did for me. We were 0–0 and I was battling hard to hold the ball up and take the pressure off the defence, the sort of situation in which I thrive. I was taken off again, which I felt made no sense, though he tried to justify it with the old standby about wanting fresh legs. Even the chairman was to question it in the paper. We lost 1–0. I was dropped for the next game against West Bromwich Albion and, when we were 2–0 down at home to Crewe, the crowd were baying for me and he had to bring me on. He had to keep me in because of injuries, but the team was on its knees. I thought I did well enough against Bolton Wanderers at home, when I hit the bar. But now Rory Allen was fit again after three months out and I just knew Bally would bring him straight back for the next game, against Sheffield United. He didn't tell me until 2p.m. on matchday, though. What was more, he told me that I could leave on a free transfer: that Lee Bradbury, who he had just signed, and Rory, were the future of the club.

I was sickened that I would have to leave the club that I had worked so hard to get back to, but in another way I was relieved. To me, he had lost it. He was holding the club back. Things got so desperate one day that we had six foreign players at a trial match on the training ground, sent by some agent for him to have a look at. They looked like Mexican goatherds except not as good. The trial was called off after 20 minutes, but that was how desperate he was becoming.

There was something called the 'experience committee' at the club, formed so that he could tell a group of the senior players things to be passed on to the youngsters. I was never on it, despite my seniority. At one point, Bally also told some of the younger players to ignore the older ones because they wouldn't be here much longer.

There was something else that topped the lot. I suppose I could have forgiven him for a lot of his offhand treatment of me, but for one thing. He was getting tips for horses and not telling me! I had to get them off Andy Awford, our club captain. Imagine that: the manager not telling me about a sure thing!

I just wanted to play again and, over the weekend, I spoke to the Brighton manager Micky Adams about a loan move. On the Monday I went in to see the managing director, David

Deacon, and told him everything that had happened. His jaw dropped. I told him that I wanted my contract paid up before I would go, and he said to leave it with him for a week and he would see what he could do. Three days later, in the first week of December, Alan Ball was sacked. I was told that the board had been stunned by what I had to tell David Deacon about the way the team was being run, but I felt that I had had to speak out in the best interests of the club. This may have played a part in his sacking, but we would have gone down, I'm sure of it, had he stayed in charge.

Under Bob McNab, the former Arsenal full-back and a friend of Milan Mandaric's in the United States, who took over as caretaker manager on the Monday, I was straight back into the side for the home game against Port Vale and I played well. Next, it was Swindon away and I scored the team's first goal for a few games. It set me off on a run of 12 goals in 16 games. As a result, the team began to clamber out of trouble and survived the First Division comfortably. The only dark point was me damaging knee ligaments at the end of the season, which meant it would be touch and go to recover in time for the start of the following season.

The appointment of Tony Pulis as manager in mid-January helped. I think Bob McNab would have liked the job, but the chairman wanted someone more in touch with the English game and Tony's experience with Bournemouth and Gillingham helped. Immediately he made a difference and we looked better organised. He is very straight and very direct, which is how he likes his teams to play, and he knows I will work hard for him. I hoped I would fit into his style of uncomplicated football.

I had never wanted to leave Portsmouth and now I was back doing what I do best, back among a group of lads who were real characters. I've been lucky in that respect, being around players who play hard and know how to enjoy themselves.

Take Jimmy Carter, the former Liverpool and Arsenal winger, at Pompey. I remember one day the physio, myself and Andy Awford were going to the bookies and Jimmy asked us for a lift to the Roundabout Hotel in Fareham. He was just wearing his training kit and didn't have any money on him. When we passed by on the way back a couple of hours later, Jimmy was sat in the middle of this busy roundabout, enjoying the sun and a couple of bottles of champagne at a table for four with a group of fellow revellers. How did he pull that off? I think Jimmy's old man had a few bob. He once turned up for training in a Bentley.

A few of the lads did get into some scrapes, it has to be admitted. In fact, if we were a Premiership club it would have been big news. Ceri Hughes was involved in a nightclub incident, for example, and on another occasion he butted Sean Derry, leaving him with a nose like Karl Malden, that bloke in *The Streets of San Francisco*. Andy Pettersen was also done for drink-driving, having smashed his car up, while Russell Hoult was fined £300 for driving a little slowly in a rather brightly lit area, the lights being red. I thought I might do something similar. It's cheaper than being fined by the FA for betting on my own team, of which more shortly. In addition, Aaron Flahavan and Rory Allen got arrested for fighting at TGI Friday's and Rory got fined two weeks' wages. The gaffer gave it to the lads for an outing to Cheltenham. In the first race I was putting it on a nag and waving the wad at Rory. "Ee are Rory, here's your wages,' I said to him, laughing. He had to take it in good part. The horse nearly fell at the first.

My own gambling was going well, a couple of the lads having

given me some good info. I had sold my old house in Luton and, with a chunk of the proceeds, I had a huge bet on Istabraq to win the Champion Hurdle, getting 10–11, 4–5 and 8–11 at various bookies before taking 8–13 on the course, before it went up to 4–9. When it came home, it meant that I would show a profit on the season. I also collected on an old bet, the bloke giving me a red Porsche as payment.

Then came the bet that landed me in trouble with the FA. I was walking into the ground one Saturday for our home game against Barnsley and called in at the on-site bookies, Flutters, to check the odds. We were 100–30 to win and I was shocked, insulted even. The bloke behind the counter just said they took the odds from the national firms and didn't set the price locally. That was some price for a two-horse race, and us on our home turf.

I went into the dressing room and told the lads, and six of us decided to have £50 each on us to win. We duly won 3–0 and I scored a hat-trick. Everyone said that I then made a mistake by talking about the bet in my post-match interviews, but there was method in my madness. It went like this . . .

Earlier in the season I had had a bet on Arsenal to beat Leicester in the FA Cup at Highbury. I had heard that Leicester were going to be without Emile Heskey, Neil Lennon and Muzzy Izzet. In fact, Leicester somehow got a 0–0 draw and I lost a bundle. I thought that if I owned up to the Barnsley bet and the FA told me to void it, then I might get my money back on the earlier, bigger bet. The firm concerned would surely have voided it, because they wouldn't want the bad publicity. That was my reasoning.

The problem was that though the FA charged me, they did not tell me to void the Barnsley bet. I even rang them up one day to check if they wanted me to, almost pleading with them, but they said there was no need. It meant I couldn't get the other bet cancelled.

I was duly summoned to London for a disciplinary hearing. At the Thistle Hotel in Lancaster Gate, they told me. The problem is that there are three Thistle hotels in that area, all within a few hundred yards of each other. Naturally, the last one I went into was the one for the hearing and I was a bit late. They were also seeing Emmanuel Petit, though, so they did that one first. The hearing, involving a disciplinary panel of three representatives from county associations, was very fair. I think they were amused rather than anything else and it was something and nothing. After all, a recent survey had shown that about 40 per cent of professionals regularly bet on the game. I was fined £900, about the size of the winnings.

What gave the case added significance was that it was heard on the day after the Hansie Kronje cricket betting scandal had erupted and the reporters outside were all asking me about gambling in sport. But there was no real comparison. I personally have never ever come across a player who would get involved in anything dubious or detrimental to his team or the game. I know for a fact that once, at Birmingham, our goalkeeper Ian Bennett was approached by some shadowy Middle Eastern figure and immediately reported it to the club. I presume they told the FA.

The arrival of spread betting did present opportunities for abuse, I have to admit. For example, you could bet on which team would have the first throw-in and it could have been tempting for a team to influence that. That's all stopped now, though. Otherwise, to me there is a big difference between betting on the opposition, which I would never do, and betting on yourself, backing your own ability. I once had a bet on Birmingham to win at Huddersfield and nobody seemed to mind that, which made being done for the Barnsley bet seem strange. But they seem to change the rules every year. One thing I did learn at my hearing was that it was now forbidden only to bet on games in your own division, but that I could bet on the Premiership or the Second and Third.

There was one losing bet during the season that I didn't pay out on. At Port Vale, we had a penalty given against us and as he was stepping up to take it, Tommy Widdrington asked me what odds I would give on him scoring. '1–2,' I said. 'I'll have a tenner on that,' he replied. I think he thought I said 2–1 and, when he scored and later tried to collect, I refused.

I am much more sensible about my gambling now, having decided that I needed to get my finances in order in the twilight of my career. I know I can never claw it all back but here is a chance to save and prosper for the few years I probably have left as a player. If my problem does resurface, I will get help. After a game for Leicester against Arsenal, I spoke to Paul Merson, who is a really good man and has shown how to deal with it.

I also think I am probably more sensible in many areas of my life. I haven't had nearly so many car accidents. There was that one where the lorry went into the back of me, leading to me having the knee operation a couple of years back, and one near Portsmouth where a bloke came out of a junction and I swerved to avoid him, but instead hit him before rolling into another car. My Rover was smashed up and I was furious with him, and he was full of apologies, saying he had been half asleep. I was even more annoyed with the woman in the third car who said she was a lawyer and was telling him to admit nothing. As for my penalty points, I managed to go two and half years on 11, so avoiding a ban, then got three more for speeding a month or two later.

In my personal life I did find that I was coming to wonder more about my background and, after talking with my mum to check that it would not upset her, I arranged to meet my blood mother, who came over from the Isle of Wight, where she works part-time in a shop. We talked for a few hours about my background and my blood father, who was apparently a good man, and she was good as gold about me regarding the people who had done so much for me, Anne and Alan, as my real parents. I also have step-brothers and -sisters and, when the time feels right for everyone in the future, I might like to meet them too.

In the end, you do acquire some wisdom to go with age and it has helped being settled back in my home town. Sooner or later you have to be taken seriously and it is hard to shake off a reputation for being eccentric, or a bad time-keeper, or whatever, though if my time-keeping ever became an issue as a manager, I would just point to Martin O'Neill, who was one of the worst at punctuality I have known, but who knows a thing or two about management, having now made it to Celtic.

I suppose I have stopped doing a lot of the daft things I used to. It's probably a sign of maturity that you start to cringe at some of the things you have done, like trying to get my feet into size fives at Bournemouth or squatting in a director's flat at Cambridge. Hopefully I haven't stopped doing all the daft things, though. You've got to enjoy yourself, haven't you? I still like a laugh and a joke, and am still part of the dressing room laughs, though probably not at the forefront of them any more. I've learned that I don't have to put myself into potentially damaging situations any more, though when I look back, it feels like they have sought me out rather than me looking for them.

Certainly some of the bizarre ones seem to have found me, like the time a while back when I went up to pick up a mate, Duncan, at the University of Greenwich, where he was a mature student. We were sat in his room having something to eat when we heard noises of moaning and laughing outside the window. When we looked out, a group of lads had a projector and were showing – in the widest of widescreen – a blue movie on the wall for the benefit of the whole college.

But just as I could be underestimated as a player because of the way I looked, so I think people who don't really know me can underestimate me as a character. I believe I am a shrewd judge of the game and clever in my own way, both on and off the field. Cleverness has nothing to do with seriousness or intelligence. After all, there aren't too many overqualified players, but I know a lot of clever ones. It's more how you can hold a conversation and amuse people. And you have to learn that in football, otherwise you go under. It is part of being a pro, giving stick and taking it. It's not just about ability. You have to have luck, but courage and personality too. You can't show weakness in a dressing room. If you are singled out, it can finish you.

Money could not buy my experiences, the characters I have met at all my clubs. I know people think that players are cosseted and put on a pedestal, that they are indulged too much, that they are stupid and don't know how to do things like open bank accounts or book holidays because they are looked after all the time, but I don't remember feeling cosseted or indulged at Aldershot or Cambridge United.

Footballers can be fragile people who get wound up, because it is such an uncertain way of life, always pressurized. I know many people also face that in their working lives and wouldn't get looked after in the same manner, but if you are going to get the best out of players it has to be that way. Football is not a normal way of life. Few will have had to put up with the fearful abuse I have. And how many people, apart from entertainers, are watched for a living? You wouldn't want to pay money to see a plumber in action, would you?

All that said, I wouldn't have swopped it. The hours may be unsocial at times but they are pretty good. I don't know of any other job, either, where I would have got the thrill that I did at Wembley for Leicester, or in scoring my first league goal against Newport for Bournemouth, and in between are my other best memories: that night for Birmingham at West Bromwich Albion, or the day we beat Huddersfield to clinch promotion. Even the worst, going down with Birmingham at Tranmere, has an intensity of feeling that few people will experience.

To the great players of the game playing in the biggest of matches some of my memories may appear small beer. But they also serve . . . I think my career illustrates the depth and colour of the English game, which sustains more full-time professional clubs than any other league in the world.

I want to stay in football as long as I can and I will run and run at whatever level I can. I would love to get back into the Premiership – with Portsmouth – but moving down through the divisions holds no fear for me. I live to play. I will keep my socks rolled down – if allowed by referees and defenders – and will chase every ball with all the enthusiasm and passion I can muster. After that, who knows? I have done some TV and radio work and enjoyed it. I reckon I have good and honest opinions and people respect that. One day, too, I would like to try my hand first at being a player-manager and then a fully fledged manager. Starting at somewhere like Weymouth or Aldershot would appeal to me. What kind would I be? Well, I wouldn't worry too much how people trained or behaved, whether they liked a drink or a bet, as long as they weren't over the top and did the business on a Saturday. No one likes to win in the game more than me but you can have fun achieving that. The club I managed would not be a boot camp, but there would be plenty of tales.

Index

Viera, Patrick 244

Waldron, Keith 240
Walford, Steve 216, 219–20, 231
Walker, Len 87, 88–9, 96, 104, 107
Walsh, Steve 219, 220, 224, 227, 228, 231
Ward, Mark 31, 174, 178, 193
Watts, Julian 227
Wealands, Jeff 76
Wealdstone 79
Webb, Neil 54
West Bromwich Albion 171–2, 250, 259
West Ham United 144, 152–4
Weymouth 9–10, 11, 23, 64, 65–7, 73–82, 83; decline of club 80; dropped for Horwich match 66–7, 81; fight for promotion to Football League 66, 78–9; first season 75; goals scored by Claridge 75; match against